Praise for *Lucky Planet*

'Very readable'

'One for the armchair speculator, [Waltham's] arguments are compelling and the book is a delight to read.'

Independent

'[*Lucky Planet* is] a lively and well argued antidote to a widespread view that advanced life could arise frequently and in many places in the known Universe. Waltham explains why the Earth is a much more peculiar planet than you might think, and he shows that its friendliness to life does not just apply to the here-and-now, but must equally have pertained through a history of more than 3.5 billion years: life's survival and prospering to the point where intelligent life could emerge was a product of extraordinary and exceptional luck. A sceptical response to ideas of inevitable evolution of intelligent beings among the stars, Waltham suggests that we may, after all, be lonelier than we could have thought ...'

Richard Fortey, author of
Survivors and *The Hidden Landscape*

'David Waltham takes us on a delightful tour of the various factors that influence planetary habitability and the evolution of advanced life. That he thinks the prospects for it are unlikely is all the more reason for us to go up to space and take a good look!'

author of
Planet

LUCKY
PLANET

DAVID WALTHAM

LUCKY PLANET

WHY EARTH IS EXCEPTIONAL – AND WHAT THAT MEANS FOR LIFE IN THE UNIVERSE

ICON

First published in the UK in 2014 by
Icon Books Ltd, Omnibus Business Centre,
39–41 North Road, London N7 9DP
email: info@iconbooks.net
www.iconbooks.net

This edition published in the UK in 2015 by Icon Books Ltd

Sold in the UK, Europe and Asia
by Faber & Faber Ltd, Bloomsbury House,
74–77 Great Russell Street,
London WC1B 3DA or their agents

Distributed in the UK, Europe and Asia

7th Floor, Infinity Tower – C, DLF Cyber City,
Gurgaon 122002, Haryana

ISBN: 978-184831-832-8

Text copyright © 2014 David Waltham

Typeset in ITC Galliard by Marie Doherty
Printed and bound in the UK by Clays Ltd, St Ives plc

Contents

About the author

David Waltham obtained a first-class degree and a PhD in Physics before moving into the oil industry in the early 1980s. This industrial experience led to his appointment, in 1986, as a lecturer at Royal Holloway, University of London, where he became Head of Earth Sciences from 2008–2012.

Prologue: A Tale of Two Planets

Far beyond the range of any telescope humanity will ever possess lies the doomed planet Nemesis. Nemesis, a near-twin to our own world, is named after the ancient Greek goddess responsible for the rebalancing of undeserved good fortune. As befits this name, Nemesis has been on a lucky roll far longer than any world could reasonably expect, but her streak of good fortune has come to an end.

Nemesis is dying. The immense herds that once swarmed across her vast plains are gone for ever. The huge beasts that swam her clear blue ocean waters are now extinct, and the previously verdant rainforests of her equatorial regions have withered and died. A beautiful and complex biosphere has vanished, leaving bacteria and a few species of worm as meagre representatives of the multifarious life-forms that once made Nemesis the biological pinnacle of her galaxy.

When Nemesis first formed she was a duplicate of the early Earth in almost every way. Most of her subsequent history, too, was strikingly similar to that of our own planet. Both of these initially sterile, over-heated hell-holes transformed into vibrant, microbe-infested globes within a few hundred million years of their births. Single-celled plants arose on both worlds within 2 billion years, giving them oxygen-rich atmospheres just as they reached their 3-billionth birthday. Breathable atmospheres in turn allowed large plants and animals to appear when Nemesis and Earth reached about 4 billion years of age. By middle-age, at 4.5 billion years, ferocious monsters stalked both worlds. Had they known it, the Earth-dinosaurs and Nemesis-dragons could have been proud of their status as the most complex organisms in their respective galaxies. Then disasters struck both planets. Only one world recovered.

The first indications of trouble on Nemesis were subtle; storms became a little bigger, droughts slightly longer and winters marginally colder. Animals and plants adapted to the changes and continued to prosper. Bit by bit, conditions worsened. During the most extreme periods the entire planet switched from stiflingly hot desert to frigid polar wasteland and back every few hundred thousand years while, at other times, much of Nemesis spent months in frozen darkness followed by months of continuous, blazing sunlight. The climate occasionally stabilised to give life a respite but, after a few tens of thousands of years, it became erratic once more and intolerable for many organisms. Eventually, plants could no longer evolve fast enough to keep up and when the plants died, so did the animals. Extinction outpaced species creation, and within a few million years of the start of the troubles, only the most robust organisms clung to life.

Earth too endured a dangerously variable climate at this time. Sulphur dioxide clouds, the result of intense volcanic eruptions, reflected sunlight into space to produce severely cold decades. When the clouds cleared, equally unforgiving hot periods followed, produced by greenhouse gases blasted into the air by those same volcanoes. And, just when it seemed that things could get no worse, Earth was hit by an asteroid. The climatic consequences of this impact proved to be the final straw for millions of struggling species; many, including the dinosaurs, died out. Yet this is where Earth's fate finally diverged from that of Nemesis. The climate chaos calmed down and Earth's biosphere began its slow journey back to full health. Our ecosystem took 10 million years to recover but new life-forms slowly emerged to repopulate a changed yet still vibrant world; a world that went on to produce *Homo sapiens*, 65 million years later.

How did two worlds enjoy billions of years of exuberant, parallel development and then eventually experience such dramatically different outcomes? There are many ways of getting a world wrong but few ways of getting it right. Nemesis stands for all the myriad failed worlds in the Universe, just one example of how not to build a habitable planet. Yet her difference from our own world is tiny, as we'll see in the pages of this book. In contrast, Earth has been blessed with incredible good fortune, giving it all the right properties to sustain a complex and beautiful biosphere. It may just be the luckiest planet in the visible Universe.

1
Almost Too Good To Be True

The Earth is a precious jewel in space possessing a rare combination of qualities that happen to make it almost perfect for life. *Lucky Planet* investigates the idea that good fortune, infrequently repeated elsewhere in the Universe, played a significant role in allowing the long-term life-friendliness of our home and shows why it is unlikely we will succeed in finding similarly complex life elsewhere in the Universe.

The proposition that the Earth may be an oddball, a planet quite unlike any other we will ever find, has been discussed for centuries. Until recently such debates were built upon mere speculation, but times are changing. We now sit at one of those scientific crossroads where a field of study moves from being a disreputable, if interesting, subject for discussion to a real science with defendable conclusions based on substantial evidence. Such transitions occur when technological advances make previously impossible observations routine and, as a result, new data becomes available.

In the case of oddball Earth, the new data comes from advances in how we look at the rocks beneath our feet and at the stars above our heads. The rocks tell a tale of our planet's constantly changing environment along with the story of life and its struggles to survive. The stars speak of many possible worlds, all unique in their own way. These parallel stories suggest that incredible good fortune was needed to allow our existence, although that proposal remains controversial. Many of my colleagues will tell you that the data are still too sparse to decide whether we live on a fairly typical planet orbiting a normal star in an unremarkable part of a common-or-garden galaxy or, alternatively, on the weirdest world in the entire visible Universe.

Personally, I no longer have doubts. The evidence points towards the Earth being a very peculiar place; perhaps the only highly-habitable planet we will ever find. This view has led some astrobiologists to describe me as 'gloomy', but I don't see things that way. For me, these ideas merely emphasise how wonderful our home is and how lucky we are to exist at all.

My central argument is based on geological evidence showing the Earth to have had a surprisingly stable climate. At first glance this may seem like a trivial claim. Why shouldn't the Earth have a stable climate? Quite simply, because the factors that control our planet's surface temperature have all changed dramatically during the 4.5 billion years of Earth's existence. Our Sun now gives off much more heat than she did when young, while geological and biological activity have produced a modern atmosphere with a completely different composition to that in the distant past. The scale of these natural variations dwarfs those imposed by mankind in the last few centuries. We have made minor adjustments to the atmospheric composition, have caused significant alterations to the amount of cloud cover, and have even destroyed entire ecosystems. Among many other nasty side-effects, our tinkering will produce a warming of the climate comparable to that experienced at the end of the last ice age. Now imagine the climatic result of atmospheric, oceanic and terrestrial changes hundreds of times bigger than those we have been able to generate. This is the scale of transformation imposed by Nature during the long history of planet Earth. And despite Nature's massive modifications, the climatic fluctuations wrought by astronomical, geological and biological processes have always more or less cancelled each other out. I find that remarkable.

There is no dispute that the Earth's climate has been continuously suitable for life for billions of years; we have incontrovertible evidence for life throughout that time. However, the reasons for the unbroken eons of life-friendly climate are hotly debated. Most scientists agree that the evolution of our beautiful, complex biosphere could never have occurred if the Earth had not enjoyed billions of years of reasonably good weather, but it is not at all clear whether there are processes that automatically stabilise our climate, and that would therefore also work on other worlds, or whether the Earth has simply been very, very lucky. It is also possible that life, once started, is more robust than we believe and would have survived even had there been more dramatic climate change over the long history of our planet. I'll consider these possibilities in the pages of *Lucky Planet*.

The obvious questions this idea raises are: 'Why should the Earth have been so lucky?' 'What's so special about us?' The answer is that we're looking at the most severe case of observational bias in the history of science. This rather sweeping statement lies at the core of my book so I'd better explain what an 'observational bias' is. Observational biases occur whenever what you see is *not* what you get. For example, on mountainsides, seashores, cliffs and other rocky places, harder rocks tend to stick out while softer ones erode away, with the resulting spaces being filled with mud, vegetation or rubble. Under these conditions it is easy to erroneously believe that the area contains only the harder rocks. *Our view of what is really there has been misled by the accident of what we're able to see.*

A similar observational bias occurs when we look at the night sky. The majority of stars visible to the naked eye are more massive than our Sun even though 95 per cent of all stars are actually lighter. The reason is simple: bigger stars are brighter stars and our unaided eyes aren't sensitive enough to see the faint ones. In addition, heavy stars are usually hot enough to shine with a white or blue light but the much cooler majority of stars would be distinctly reddish if we could only see them. The few thousand stars we see on a dark night are therefore unrepresentative of the hundreds of thousands of stars that inhabit our small corner of the galaxy. To eyes that could see these red-dwarf stars, the heavens would be awash with faint red points of light interspersed only rarely by the brighter white stars, blue stars and red giant stars that dominate the night skies seen with human eyes. *Our view of what is really there has been misled by the accident of what we're able to see.*

The potential for observational bias becomes enormous when the Earth itself is the subject of enquiry. In the same way that we can't see rocks that are buried, or stars that are faint, intelligent observers can't see a home-world that is uninhabitable. We must be living on a planet suitable for intelligent life, even if such worlds are extraordinarily rare and peculiar. As a geologist I think this 'anthropic selection effect', as it is known, is a vital but almost universally ignored insight and we simply cannot understand our planet properly without taking it into account. *Our view of what is really there has been misled by the accident of what we're able to see.*

As a consequence of this bias, we must acknowledge and take account of our privileged viewpoint when considering whether qualities of the Earth are typical or exceptional. An instructive example concerns the surprisingly early appearance of life on our planet. The fact that microbes appeared on Earth while our world was still very young is often taken as evidence that life appears easily and will be widespread throughout the Universe. This is mistaken. Planets are habitable for only a few billion years and so intelligent life probably doesn't have time to evolve on worlds that drag their feet over life's origin. All intelligent observers, including us, must find themselves looking out onto worlds where life began soon after conditions became suitable. The *possibility* that this is a chance event not repeated on most habitable worlds means that there *could* be an observational bias and an early start for life on Earth cannot be used as evidence that life is an easy trick for a planet to pull off. Maybe it is and maybe it isn't.

From my perspective, the most important anthropic selection effect concerns the resilience of life. I've frequently heard it said that life is exceptionally robust, once it arises, as shown by the fact that it has survived every catastrophe thrown at it during Earth's long history. But how could it be otherwise? Planets where life fails to survive do not give rise to sentient beings. Intelligent observers throughout the Universe, no matter how rare or common they may be, must look out onto home planets where life has managed to survive. Perhaps life doesn't survive for long on the majority of planets where it appears and we simply wouldn't be around to notice had the Earth been less fortunate.

A planet may therefore have to be pretty weird to allow a creature as odd as *Homo sapiens* to appear. However, for practical reasons, *Lucky Planet* will discuss the planetary preconditions necessary for complex life-forms in general rather than sentient life-forms in particular. From observations of the Earth's biosphere we can say a great deal about the environments that favour complex organisms, but it is much harder to say anything concrete about the circumstances under which intelligence emerges.

Given this generalisation, I should be clearer about what I mean by complex life. In the case of Earth life it is helpful to draw a distinction between single-celled organisms and multi-celled ones. The vast majority of organisms on this planet are microscopic, single-celled creatures such as amoebas and bacteria. These are anything but simple. However, some rather rare organisms have relatively recently evolved the trick of growing enormous colonies of cells tens of metres tall (e.g. trees) while their close relatives have evolved similarly large colonies able to move about to track down food (e.g. grazing cows). These multi-celled organisms have an even higher level of organisation than their single-celled relatives. Single-celled organisms do sometimes form colonies but the key characteristic of more complex organisms is that they are constructed from many different types of cell. Of course, we shouldn't be too Earth-centric in our thinking. Perhaps complex life-forms on other planets are not multi-celled creatures with differentiated tissues but have a completely alien and utterly unimaginable architecture instead. Nevertheless, I think we can be sure that alien intelligences, if they exist at all, will be more complex than single-celled Earth organisms. As I'll show in later chapters, simpler organisms tend to be much tougher than more complex ones and so this distinction is quite important.

Lucky Planet is an exploration of the idea that the Earth is a very strange place – perhaps the luckiest planet in the visible Universe. We'll begin with the opposite idea, the scientifically conventional one that there is nothing particularly special about our world at all. We will then tour astronomy, geology, climatology, biology and cosmology to show why this conventional view needs to be reconsidered. In many places you will almost certainly come up with counter-arguments. However, I hope you will still conclude that 'Is the Earth special?' is a sensible question to ask. After this tour, I'll return to Nemesis, the 'unlucky planet' with which I began. Once you see how trivial the difference was between Earth and our near-twin, I hope you will agree that our planet really is almost too good to be true.

2
Mediocrity

On St Valentine's Day 1990 the voyager turned around for a last lingering look at the home she had left twelve years before. Thus, from beyond the orbit of Pluto and 6 billion kilometres from Earth, the space probe Voyager 1 took the most distant photograph of our world ever attempted. The result was a picture of an insignificant spot barely visible against a background of instrument-scattered sunlight, but this picture beautifully encapsulates our modern view of the Earth as a tiny, unimportant speck in space. Carl Sagan, one of the greatest-ever popularisers of science and the man who did most to encourage NASA to turn Voyager 1 around to capture this image, memorably described the Earth in this picture as just a 'pale blue dot'. At that time we did not know whether stars, other than the Sun, had planets. It was still possible to believe that there was something special about our star's entourage of six pale, variegated dots (Venus, Earth, Jupiter, Saturn, Uranus and Neptune had all been imaged) but this state of affairs didn't last long. The first widely accepted exoplanet, a planet orbiting another star, was discovered just five years after Voyager 1's farewell photograph and, by early in the 21st century, it has become clear that exoplanets are pretty common. These discoveries, along with Voyager's photo, reinforce the perception of our world as small, insignificant and lost in the immensity of the Universe. In this book I plan to challenge that view and, to begin with, I want to look at the historical background to the idea that the Earth is a mediocre planet.

The idea that our world is just one planet among many has certainly not been mankind's view through most of history.

Until 400 years ago we generally placed the Earth at the centre of the Universe or, with a little less hubris, at the bottom of the ladder to the heavens. The first step towards an improved sense of perspective was taken by Nicolaus Copernicus, whose *De Revolutionibus Orbium Coelestium* (On the Revolutions of the Heavenly Spheres) was posthumously published in 1543. This book revolutionised our view of the Universe by suggesting that the Earth and planets revolved around the Sun rather than all heavenly bodies revolving around a stationary Earth. Interestingly, *De Revolutionibus'* title is the origin of 'revolution' as a word to indicate overthrowing of previously well-established ideas or organisations. It was another 150 years before this first revolution, the Copernican revolution, became widely accepted. Nevertheless, once the Earth had been knocked off its perch at the centre of the Universe, the next step in our planet's demotion came along with remarkable rapidity: perhaps the Sun isn't the centre of the Universe either!

Giordano Bruno, a 16th-century priest, was among the first to wonder whether the stars are just distant suns and whether these too have planets revolving around them. The story is that Bruno was burned at the stake for suggesting this and for supporting the views of Copernicus. Bruno is therefore held up as an early scientific martyr, someone who gave his life in the battle of truth against ignorance. However, this tale of scientific heroism is a 19th-century exaggeration promoted at a time when the supporters of Darwin's new theory of evolution saw themselves in conflict with a church they regarded as superstitious and reactionary. The myth was magnified further by the resonance it had in a 19th-century Italy struggling to emerge as a nation and straining to free itself from the political dominance of the Vatican. As part of the propaganda campaign in this power struggle, a statue of Bruno was erected in 1889 near to the spot where he was executed, further fuelling the secular canonisation of this controversial and colourful character.

Thus, 250 years after his death, Bruno was dragged into two new revolutions: one that began with the 'Spring of Nations' nationalist uprisings of 1848 and one that began in 1859 with the publication of Charles Darwin's *On the Origin of Species*. However, the dreadful fate of Giordano Bruno was the consequence of a much earlier and even bloodier clash of ideas.

At a time of great religious strife in Europe resulting from the rise of Protestantism, the rather argumentative Bruno travelled through Italy, France, England, Bohemia and the Germanic countries and, as he travelled, he argued with almost everyone about almost everything. To his credit, he was trying to reunite a divided Western Europe behind his own 'Hermetic' version of Christianity. Hermeticism has existed in one form or another since the early years of the Roman Empire and still has its supporters in our own time. Over that 2,000-year history it has meant many things to many people but one of its more constant messages is that everything is divine, even the rocks of the Earth. The idea that all of creation is sacred contrasted starkly with the religious orthodoxy of 16th-century Europe, which held that everything below the orbit of the Moon, the sub-lunar world, was degenerate, while the heavens beyond were incorruptible, eternal and perfect. Despite this rather serious barrier to wide acceptance, Bruno saw Hermeticism as a way to bridge the theological divide between Catholics and Protestants and, at a time when many were tiring of bloodshed, his ideas might have been listened to but for his rather arrogant, tactless and belligerent manner. Instead, he merely succeeded in angering all sides and even managed to be simultaneously excommunicated by the Calvinist, Lutheran and Catholic churches; almost the full set. Then he suffered the possibly worse fate of being laughed at in Oxford, an experience he neither forgot nor forgave.

Despite his Oxford experiences Bruno remained in England for two years and, during this visit, made the fatal mistake of becoming embroiled in the intrigues of the French ambassador against the Spanish by acting as a spy (as well as, possibly, a double agent spying on the French for Queen Elizabeth). I'll come back to the fatal consequences of this unwise move later, but it was also while in England that Bruno wrote *La Cena de la Ceneri* (The Ash Wednesday Supper) and *De l'Infinito Universo et Mondi* (On the Infinite Universe and Worlds), in which he explained his cosmological ideas. In these books Bruno took Copernicus' suggestion that the Earth went around the Sun to its logical conclusion: if the Earth moved just like the other planets, then the Earth and the heavens were not fundamentally different. Clearly this fitted well with Bruno's belief that the Earth and heavens were equally sacred and it explains why he supported Copernicanism so strongly. Bruno then took his ideas a breathtaking step further by reasoning that, if the heavens were made of the same stuff as the Earth, there was no reason why there could not be Earth-like places elsewhere in the heavens. Perhaps the other planets are just like the Earth and perhaps even the stars are just distant suns each with their own planetary companions. Here, Bruno was expressing for the first time what we now call the principle of mediocrity – the idea that there is nothing special about the Earth. The Earth is just a typical planet in orbit around a typical star.

Less than ten years after Bruno's execution his compatriot, Galileo Galilei, became one of the first to turn a telescope onto the night sky and what he saw proved beyond all reasonable doubt that Copernicus was right – the Universe did not revolve around the Earth. Galileo was the first to see that Venus showed phases like the Moon, phases whose timing made sense only if Venus went around the Sun and not around the Earth. He also saw that Jupiter had its own moons and this again showed that the Earth did not lie at the centre of all things. Finally, he saw that Earth's satellite is an entire world complete with its own mountains, valleys, cliffs and craters. It still took decades to convince those unwilling to accept the evidence of their own eyes but Galileo's observations proved that Bruno had been right: the Earth is not the only world.

Over the next 400 years, the Earth was demoted further as we discovered that the Sun is just one star in 200 billion forming our galaxy. Our galaxy, in turn, is just one out of hundreds of billions of galaxies in the visible Universe. And it is probable that the visible Universe is only a small fraction of the entire Universe, with recent speculations even suggesting that the Big Bang was a local affair and that our Universe is just a tiny part of what some are calling a multiverse, a subject I'll return to much later. Furthermore, the last few hundred years of research have shown that the principles of physics and chemistry are the same in the most distant parts of the visible Universe as they are on Earth. The conclusion then, following centuries of scientific work, is that the Earth is nothing special and its location is very ordinary.

Assuming the Earth to be mediocre has been a powerful tool enabling us to greatly expand our horizons and to see the true vastness and grandeur of the cosmos. The principle of mediocrity has served us well for nearly half a millennium but I believe that its very success has caused this invaluable working principle to slowly mutate into an unbreakable law. Ironically, it has become scientific heresy to question Bruno's insight. An almost subconscious belief in the ordinariness of our world is making us blind to an important truth: there may be things about our planet that are far from typical. As I've already discussed, places suitable for the emergence of intelligent observers may be extremely rare. We might therefore need to return to a geocentric cosmology in the sense that the Earth may be the most interesting place in the observable Universe.

Before moving on, I'd like to complete Bruno's tragic story. Several years after leaving England he returned to Italy and worked for the nobleman Zuan Mocenigo in Venice.

The Catholic church already considered Bruno to be a dangerously unorthodox thinker but he should have been safe in a Venice that was proudly independent of papal influence. Unfortunately he angered Mocenigo by refusing to teach him black magic. His entirely reasonable excuses were that, despite rumours to the contrary, he didn't know any magic and didn't approve of such things anyway. However, he must have said this with all of his usual tact and diplomacy because Mocenigo took offence and denounced Bruno to the Venetian Inquisition. The Venetian Inquisition, in turn, passed him on to the Roman authorities and Bruno was snared.

As was the usual fate of heretics at this time, the Roman Inquisition locked Bruno up and threw away the key. He probably expected to spend the last few decades of his life in jail, but events in Spain and southern Italy led him to an even worse fate. A revolt broke out in Spanish-ruled Calabria and the leader of the revolt just happened to be another Hermetic philosopher. After suppressing this revolt, the Spanish authorities decided they wanted to make an example of Bruno, the most famous Hermetic philosopher in Europe as well as someone who had plotted against Spanish interests while in England. Spain therefore demanded that Bruno be executed. Rome, in turn, was looking for favours from the Spanish and so, on 17 February 1600, Bruno was led from his cell, had his tongue spiked to silence him for ever and was burned at the stake without the usual consolation of being strangled first.

It's clear that Bruno was more a victim of political circumstance than a martyr to science. Indeed many of his ideas, and his reasons for supporting them, seem distinctly unscientific today. However, we all like to have heroes, and in the 400 years since these events, Giordano Bruno has been transformed into a free-thinker whose ideas were centuries ahead of his time. There is much truth in this, even though the details show him as frequently quarrelsome and only occasionally profound. But, when it comes to stars being suns each with their own systems of worlds, Giordano Bruno hit the nail on the head and started a revolution in thought that continues to this day.

As often happens to new ideas, the principle of mediocrity built up a bit of momentum before it became widely accepted and, as a result, its eventual triumph led to a dramatic switch from outright rejection to over-application. The three centuries following Bruno's death were characterised by almost unquestioning certainty that the worlds of our solar system are fundamentally so similar to the Earth that they must all be populated by intelligent life. Ironically, this became the new religious orthodoxy since many thinkers could not understand the purpose of other worlds unless God had placed people on them.

However, a few dissenters did question this new doctrine. Of particular note is the mid-19th-century polymath William Whewell, the man who coined the term 'scientist' and a great thinker who made innovative advances in fields from geology to mathematics. Whewell is chiefly remembered today for writing an influential book on natural theology (the idea that nature's 'perfections' demonstrate the existence of God), which was referred to by Darwin at the beginning of *On the Origin of Species*. At this point in my story, though, *Origin* lay in the future. In 1853, six years before publication of Darwin's masterpiece, Whewell published *Of the Plurality of Worlds* in which he attempted to demonstrate that the Earth is special and that life is unlikely elsewhere in the cosmos. Whewell's motivation was explicitly religious: he believed the existence of intelligent life on other worlds to be incompatible with mankind's special relationship to God. But despite this, his core argument was pure science. Whewell used the new geological knowledge of his time to show that, for the vast majority of its history, the Earth had been a planet without sentient life. Hence, we have only to look at the ground beneath our feet to see that worlds uninhabited by people are logically possible.

In many ways Whewell's arguments from 150 years ago are strikingly similar to some that will be put forward in these pages. For example, I can only applaud his statement that 'the history of the world, and its place in the universe, are far more clearly learnt from geology than from astronomy', although it should be admitted that, as I'll discuss in the next chapter, that situation is now changing rapidly. Furthermore, Whewell expresses my own views perfectly when he writes that 'the Earth, then, it would seem, is the abode of life ... because the Earth is fitted to be so, by a curious and complex combination of properties'. I was also particularly struck on reading that 'the Earth's orbit is the Temperate Zone of the Solar System ... [since] the Inner Planets bear no infrastructure of life; for all life would be scorched away along with water, its first element'. This may be the first-ever reference to what is now called the habitable zone; the zone around a star where temperatures allow the existence of liquid water. There is, however, one fundamental difference between Whewell's book and my own. Whewell believed our good fortune in living on a well regulated world to be the result of divine providence, whereas I put it down to good fortune: a good fortune that is inevitable somewhere in a big enough universe.

Whewell aside, the 300 years from Bruno until the beginning of the 20th century were characterised by almost universal acceptance of the idea that intelligent life exists throughout the solar system and beyond. During the course of the 20th century, however, this came to look more than a little optimistic. The debate over whether there is life on Mars illustrates that century's transition to pessimism particularly well and so it's worth taking a look at that in some detail.

The story begins with yet another great Italian, one who lived 250 years after Bruno and Galileo. In 1877 Giovanni Schiaparelli made ground-breaking observations of Mars at a time when it was unusually close to us. Earth and Mars approach one another every two years; Mars circulates around the Sun once in the time it takes Earth to go around the Sun twice. However, the orbits of Earth and Mars are not perfectly circular and so their exact distance apart varies from one opposition (as the moment of closest approach is called) to another. In the recent 2003 opposition, for example, Mars was closer to us than it has been for 60,000 years. Schiaparelli used his drawings from the almost equally good 1877 opposition to create a new map of Mars that was significantly better than anything previously produced. Indeed, the map was so good that many of the names he chose for features still appear on modern maps based on space probe pictures. The high quality of Schiaparelli's work shouldn't surprise us. When he began his Martian work Schiaparelli had been chief astronomer at Milan's observatory for fifteen years and his most important contributions to astronomy until then had been his discovery of the asteroid Hesperia in 1861 and his 1866 demonstration that comets generate meteor showers. He also made sophisticated breakthroughs in the mathematics of orbit calculation. It's clear that Schiaparelli was an extraordinarily accomplished and talented man, but today he is chiefly remembered for a mistake he made on that, otherwise excellent, map of Mars.

Conditions for viewing at even the best-situated observatories are variable, but during those moments of exceptional atmospheric clarity that arise for a few seconds every now and again on the best nights, Schiaparelli glimpsed long, thin, dark lines crossing the surface of Mars. He dutifully marked these onto his map and called them *canali*. Many other respected astronomers confirmed his discovery and their existence was rapidly accepted by everyone. Then the trouble started. The word 'canali' was perfectly correctly translated into English as 'canals', but while the Italian word does not necessarily imply an artificial channel, the English translation certainly does. This was the origin of the myth, popular through much of the 20th century, that intelligent Martians are living on an old, dying world and have constructed a network of canals for transporting water from the poles towards an arid equator. This vision of 'intellects vast and cool and unsympathetic, [who] regarded this earth with envious eyes', was immortalised in *The War of the Worlds* in 1898 and H.G. Wells's story of a Martian invasion remained plausible throughout the first half of the 20th century. However, by the time of a major Hollywood film version in 2005, the moviemakers were distinctly vague about where exactly the invaders came from. By the early years of the 21st century, Mars had become an unlikely site for complex life. Why?

Arguments over the origin of Martian canals began almost as soon as Schiaparelli announced his discovery. The chief proponent of the view that they had to be artificial was Percival Lowell who, in 1894, established one of the best observatories in the world in Flagstaff, Arizona, with the specific objective of investigating the non-natural features of Mars. Lowell published *Mars and Its Canals* in 1905 and his book included detailed drawings showing 400 canals that were geometrically straight, thousands of miles long and frequently double 'like the rails of a railway track' as Lowell described them. For Lowell the implications were incontrovertible; the canals were artificial.

A riposte came quickly and from one of the most venerable scientists of the day, Alfred Russel Wallace. Wallace has a slightly mixed reputation today. He is revered by biologists as the co-discoverer of the theory of evolution by natural selection and as the father of biogeography, the study of the influence of geography on the distribution of species. On the less positive side, he is also remembered for his belief in spiritualism and for his campaign against smallpox vaccination. The story of his co-discovery of natural selection is well known and widely discussed in other books; in brief, Wallace formulated the theory while working in Indonesia and sent an account of it to London, where it was forwarded to Charles Darwin. Darwin, who had arrived at the same theory years earlier but had not yet published it, was stung into action and the result was a joint presentation to the Linnaean Society in 1858 followed by publication of Darwin's *Origin of Species* in 1859. Wallace's own views concerning Darwin's primacy in this discovery can be judged from the fact that Wallace published a book called *Darwinism* in 1889. This is testament both to the generosity of Wallace's nature and to the strength and breadth of the stunningly detailed evidence that Darwin had amassed for their theory during the years before Wallace's bombshell letter arrived from Indonesia.

In 1907, almost 50 years after this central role in one of the biggest scientific revolutions of all time, Wallace published *Is Mars Habitable?*, a short book that is almost a line-by-line demolition of Lowell's *Mars and Its Canals*. At the age of 83, Wallace still had a keen mind and an ability to cut through to the key issues. He did not question the reality of the canals, since they had been seen by many highly skilled observers, but he did focus on the major error in Lowell's arguments: his assertion that the Martian climate is warm enough to allow the presence of liquid water. Wallace and Lowell were both aware that the mean temperature of a planet depends on the amount of heat it receives from the Sun, the fraction of that heat absorbed rather than reflected and, finally, on the strength of the greenhouse effect. Lowell's estimates of these factors yielded a mean Martian temperature of 9°C while Wallace, after consulting the eminent physicist John Henry Poynting, arrived at an estimate of −38°C. Our modern estimate is an even colder −55°C, a figure that is now, of course, supported by direct measurements from spacecraft on the surface of Mars (in a pleasing coincidence, just as I wrote that last sentence I learned that the latest of these, the Mars Science Laboratory rover Curiosity, landed successfully three hours ago in Gale Crater on Mars).

Interestingly, Lowell's temperature estimate was wildly incorrect for almost exactly the same reason that some modern climate-change sceptics overstate the resilience of Earth's climate: he over-estimated the cooling effect of cloud reflectivity relative to the greenhouse warming effect of atmospheric water vapour. As a result, he believed that the large amount of water on the Earth made it relatively cool and that the much drier Mars would be almost as warm despite being further from the Sun. In reality, and as Wallace discussed in great detail, Mars was far too cold to allow Lowell's concept of Martian snow melting at the poles each summer, or the flow of water in canals across its surface. Indeed, as Wallace strongly suspected and we now know, the apparent shrinking of the Martian polar caps each summer results from thawing of a thin carbon dioxide frost and not from melting of water at all.

Despite the work of Wallace and other sceptics, the canal myth was finally buried for good only in 1965 when the US space probe Mariner 4 passed by the planet and sent back the first close-up pictures. Canals were nowhere to be seen, neither in the Mariner pictures nor in any of the detailed pictures sent back by the numerous space probes that have visited Mars since that time. I have spent several surprisingly enjoyable hours staring at the maps made by Schiaparelli and Lowell, along with photographs from the Hubble telescope and images from spacecraft now orbiting Mars, to see if any of the proposed canals are real features. I have even deliberately blurred the modern pictures to try to recreate the difficulties of seeing Mars through Earth's atmosphere. My impression is that a few of Lowell's canals and the 'oases' that formed at their intersections do correspond to chance alignments of features such as large craters and volcanoes and, with the eye of faith, even with some of the very largest drainage features such as Eos Chasma. However, these possible correlations are extremely speculative and, when it comes to Schiaparelli's map, I'm afraid I have been unable to explain any of his canal locations at all! Sadly, the Martian canals were a classic example of the human mind's extraordinary ability to construct patterns from the flimsiest of evidence, a useful trait but one that easily fools us into seeing things that just aren't there. The canals were 'seen' by many eminent astronomers but they are no more real than the constellations we see in the sky or ley-lines (supposed 'lines of power' constructed by imaginative individuals that connect ancient monuments on Earth). The human eye and brain just like to connect dots. Canali do live on, but on a different planet: Venus, where enigmatic but undoubtedly natural channels thousands of kilometres long have been found and that have been termed canali in, I hope, a respectful gesture to Schiaparelli. But, with the dismantling of the myth of Martian canals, the remaining slim evidence for intelligent life on Mars evaporated.

Since 1965 Mars has been visited by many spacecraft that have sent back incredibly detailed pictures of a not entirely alien world. A beautiful planet is revealed with volcanoes, canyons, plains and ice caps. It could almost be home except that there are no plant-covered continents and no blue seas. In fact, there is no stable liquid water at all because the surface is too cold and the atmosphere too thin. But something cut those canyons! Where has the water gone? The real geography of Mars demonstrates beautifully that there are no guarantees of climate stability. Mars once had liquid water and, possibly, warm seas on its surface. Sadly, the planet slowly lost much of its water to space, and the remainder became locked up in ice at the poles and below the surface. Lowell's vision wasn't entirely wrong. Mars really has turned from a water-rich world into a desert planet, but this happened billions of years ago and no sentient life-forms were there to rescue the planet by constructing a worldwide web of irrigation.

Despite the unpromising Martian environment now revealed to us by probes and landers, the debate over life on

Mars continues to the present day – although we are now resigned to searching for past or present microbial life buried deep beneath the frigid, dry and caustic soil forming the surface. The idea of simple life on Mars is far from implausible and there is tentative evidence that life may have existed there 3.6 billion years ago. In December 1984 a meteorite was picked up in the Alan Hills area of Antarctica and taken back to the USA for study. The meteorite was labelled ALH84001 to indicate its provenance and discovery date but little else was done to it for several years. When it was looked at more carefully in the 1990s it proved to be the most significant meteorite to hit the Earth since the one that may have killed the dinosaurs. ALH84001 is a meteorite made from 3.6 billion-year-old Martian rocks and it appears to contain biochemicals along with fossil micro-organisms. This meteorite therefore suggests there is a plausible case for life on Mars, but how good a case is it?

The first part of this claim states that ALH84001 comes from Mars – but how could we possibly know that? Very large meteorites occasionally hit all planets and some material from the struck world is blasted into space when this happens. Eventually, these fragments may meet another planet and so, over millions of years, significant amounts of rock are exchanged between worlds. One interesting consequence is that Earth and Mars are not biologically isolated from each other. Rock fragments blasted from the Earth will contain bacteria that could survive the journey from one world to another, provided the fragments are large enough and do not spend too long in space. So, even if life really does exist on Mars, it may well be trans-planted Earth life. It's even possible that Earth life originally came from Mars!

As a result of this exchange of rocks, there must be many pieces of Venus, the Moon and Mars sitting on the Earth; the problem is in recognising them. However, since we have been to the Moon and have sent measuring instruments to Mars, we know something of the chemical composition of these worlds, which means that we can recognise chunks of the Moon or Mars when they are found here. Chips off the Martian block are pretty rare, though. Of the tens of thousands of meteorites that have ever been collected, only about 100 come from Mars. It is clear that these meteorites share a common origin because many aspects of their chemical compositions are similar. We also know that one of them, called EETA79001, came from Mars because it contains bubbles of gas with a composition identical to that of the Martian atmosphere as measured by NASA's Viking landers in the mid-1970s. So, the claim that ALH84001 is a bit of Mars is pretty sound and not seriously disputed by any of the experts.

The second claim, that ALH84001 contains life-like chemical traces, is more contentious. The fractures in this meteorite contain iron-based minerals and complex carbon compounds that, on Earth, would definitely indicate the presence of bacteria. However, we cannot be sure that these minerals and compounds are only ever produced by living organisms. It is possible that they could be produced on Mars by exotic, non-biological, chemical processes unfamiliar to us on an Earth whose organic chemistry is utterly dominated by life. It is also possible that these biomarkers, as chemical fingerprints of life are sometimes called, resulted from contamination by Earth bacteria after the meteorite landed on the Antarctic ice sheet. So the biomarker evidence is interesting but not conclusive.

The final and most dramatic claim is that there are microscopic fossils within ALH84001. Electron-microscope images show worm-like features that certainly look bacterial but they are tiny compared to Earth-born microbes. The NASA team that made these discoveries worked hard to eliminate the possibility that these microfossils resulted from contamination or that so-called artefacts were created accidentally when the specimens were coated with gold and palladium (a standard procedure in electron-microscopy). Despite these precautions, most plan-etary scientists continue to believe that these creatures either crawled in after the meteorite fell to Earth or that they are tiny, artificially created, pellets of gold or palladium that just happen to look a bit bug-like.

So, the evidence for life on Mars in the distant past is worth taking seriously, but most experts remain unconvinced. The reason for continued scepticism is the widely accepted principle that extraordinary statements require extraordinary evidence. The evidence for life in ALH84001 would be uncontentious, not to say uninteresting, if this was a rock from somewhere on Earth – but it's not from Earth, it's from Mars. If the evidence is taken at face value and we all accept that there is, or at least has been, life on Mars, we might well discover later that under-standable enthusiasm has clouded our judgement.

If it does turn out that ALH84001 has been over-optimistically interpreted, it wouldn't be the first time. Throughout the modern scientific era there has been a tendency to over-optimism on the question of whether life in general, and intelligent life in particular, is common in the Universe. Of course, it doesn't necessarily follow that present optimism concerning life around other stars is misplaced but it should give us pause for thought. We've seen that many of the argu-ments in favour of a Universe with widespread life have been based on wishful thinking rather than hard science. Indeed, at times, it has almost been seen as reprehensible to believe that we are alone. This view, that there is something morally wrong in regarding the Earth as special, persists to this day but now takes the form of accusations of arrogance against those of us who think Earth-like biological richness may be rare. Dictionaries define arrogance as a 'display of superiority or self-importance' and so, strictly speaking, it really is arrogant to assume that the Earth is special – but that doesn't make the view incorrect. It's possible that I am both arrogant and correct!

A related complaint against such ideas is that they are 'Panglossian'. The fictional Dr Pangloss, in Voltaire's *Candide*, was a figure of fun. His eternal optimism that we live on the best of all possible worlds was most famously illustrated by his claim that we have noses so that we have somewhere to place our spectacles. The idea that our planet may have unusual features that make it particularly well suited to life is sometimes dismissed as a similarly naive belief that everything in the world is as good as it could possibly be. However, this characterisation is far from accurate since our planet is certainly less ideal for life than it could be. The reason I can state this so confidently is because 'perfect' planets are likely to be much less common than 'good enough' planets. The average inhabited planet is therefore neither as good as the best of all possible worlds nor as bad as a typical world. Instead, most intelligent beings in the Universe (including us) look out on to worlds that are neither ill-suited to life (and common) nor ideal (and vanishingly rare). We probably live on an extremely rare, but not vanishingly rare, second-best of all possible worlds.

More generally, the anthropic principle itself is deeply disliked by many scientists because they fear it encourages lazy thinking. Why bother trying to find out the real reason why the Earth has particular attributes if you can just dismiss them as mere accidents that happen to be essential for our existence? In one scientific paper I read recently, such thinking was dismissed as 'scientifically unsatisfying'. Again, this may be a fair criticism but it has no bearing on whether Earth really does have an unusual combination of properties that make it peculiarly well suited for life. In a similar vein, I have had correspondence with one deep thinker on the topic of the anthropic principle who, among the many other serious issues he raises, worries that it encourages a tendency towards atheism. My flippant response to this is that we shouldn't tell God how to run the Universe! And while I do think my correspondent has a point, I don't think it is relevant to the question of whether or not the anthropic principle is true. The anthropic principle may well encourage arrogant, lazy, naive and atheistic thinking but these anxieties concern the moral status of the theory, not its truth status.

These types of criticism remind me of similar ones made in the 19th century against Darwinism. There is a widely repeated story that a Victorian lady responded to news of Darwin's ideas with: 'Descended from the apes! My dear, we will hope it is not true. But if it is, let us pray that it may not become generally known.' This apocryphal tale illustrates nicely the moral difficulties that 19th-century people had with accepting the theory of evolution by natural selection. Indeed, it could be argued that their concerns were vindicated; there really is a plausible link between *Origin of Species* and *Kristallnacht* but this no more falsifies Darwin's ideas than the *Enola Gay* falsifies those of Einstein. Similarly, there are many excellent reasons for disapproving of the anthropic principle but none of them has any bearing whatsoever on whether it is true. Instead, we need to look fearlessly at the evidence and go with what that tells us, whether we like it or not.

In my view, the principle of mediocrity has become too much of an article of faith and the time has come when it should be questioned. Maybe there are at least some things that are special and different about the Earth compared to the vast majority of planets. On the other hand, we shouldn't just throw mediocrity away. It has proved far too valuable over the centuries for that. Clearly, there is a conflict between a logically unassailable anthropic principle and an observationally well supported principle of mediocrity, but this tension can be resolved by combining them into a single principle: the Earth is no more special than it has to be to allow the existence of intelligent life. To put it another way, the Earth is probably a fairly typical inhabited planet since, by definition, we are more likely to be living on a typical inhabited world than on an atypical one. This principle of anthropic mediocrity answers a criticism I frequently hear: that the special conditions I claim are needed on Earth apply only to Earth-like life, and that perhaps other forms of life are possible and are less picky. However, this leaves me wondering: Why should Earth have such odd forms of life if other types of life are more common?

The conventional view discussed in this chapter, that the Earth is a fairly typical world, is challenged in the next chapter, which looks at how the latest discoveries concerning other worlds are overturning this centuries-old idea.

3

Rarely Earth

Earth-like worlds are extraordinarily rare but there are vast numbers of them. That is not a contradiction! Even if worlds as highly habitable as our own are so rare that they are typically separated by billions of light years, that is a tiny step compared to the immensity of our Universe. The existence of vast numbers of near-twins to the Earth in really deep space therefore remains all but inevitable. That's how I know that Nemesis, or something very like it, almost certainly exists somewhere in the cosmos. This chapter looks at these points in more detail.

One reason Earth-like planets are probably rare is that there are just so many different ways to make a world. Until recently, we greatly underestimated the variety of worlds; but the more we look, the more varied and delightful our cosmos becomes. From the moons of our solar system through to planets circling distant stars, late 20th- and early 21st-century technology has replaced the imaginative fantasies of earlier generations with real, idiosyncratic worlds of previously unimagined variety and splendour. It could be argued that we have been spoiled by all this richness. For example, my university department runs a planetary geology course and, as part of this, we set up telescopes for our students to look at the stars and planets for themselves. With the exceptions of the Moon and Saturn, which always cause a sharp intake of breath when first viewed telescopically, some of our novice stargazers are a little disappointed by the experience. Partly this is because of the limitations of our relatively small telescopes and the quite appalling light pollution in the south-east of England. Mostly, however, it's because the beautiful images from space probes and from the Hubble Space Telescope have produced unrealistically high expectations. There is nothing quite like seeing things with your own eyes, but, nevertheless, I have to reluctantly admit that a web browser is a more useful tool for studying planetary geology than a telescope. Indeed, I'd recommend surfing the web for images of other worlds as you read this chapter!

For me, the realisation that nothing I could see through a telescope would ever again match the achievements of the space age came when Pioneer 10 flew past Jupiter in December 1973. In those pre-internet days it took months for high-quality pictures to appear in magazines so I'm not sure when exactly in 1974 I first saw them, but I do know that they made a deep impression on me. Pioneer 10's pictures have since been far surpassed, but in the early 1970s no one had yet seen anything quite like those images of an agitated alien world. For the first time we could see the intricate, colourful and ever-changing patterns of the Jovian weather. Continent-sized ripples and spots looped and curled along violently shearing boundaries between alternating orange and white cloud-belts and ruddy eddies churned inside the centuries-old storm of the Great Red Spot.

Pioneer 10 also took the first space probe pictures of Jupiter's moons but they were too distant for the images to reveal much. Better photographs had to wait for missions that followed in Pioneer's footsteps over the next few decades. When they came, the results were astounding. The Jovian satellites form a solar system in miniature with 67 moons, at the last count, orbiting at up to 30 million kilometres from Jupiter. Most of the moons are tiny lumps of rock and ice just a few kilometres across but the biggest exceptions, literally, are the four satellites discovered in 1610 by Galileo and mentioned in the last chapter. These moons were probably the first new worlds to be discovered in at least several hundred thousand years, since our earliest ancient ancestors had become aware of Mercury, Venus, Mars, Jupiter and Saturn. It is hard now for us to imagine the shock, wonder and disbelief that Galileo's revelation must have caused.

In the three-and-a-half centuries following Galileo's discovery, astronomers found that three of the Galilean satellites are bigger than our own Moon while the fourth is only slightly smaller. Despite their large size, nothing much else could be made of these worlds from Earth-based telescopes sitting more than 600 million kilometres away and, as a consequence, pre-space-age ideas about their appearance were largely guesswork. It was realised that they differ in size, density and brightness but the expectation was that they would all be dead, dusty and pocked with ancient craters just like our own Moon. Voyager 1, which reached Jupiter in 1979, showed that nothing could be further from the truth. The Galilean moons are astonishingly different from each other and quite unlike our own satellite. Their surprising and diverse nature was reinforced by the subsequent Galileo mission (named of course after the moons' discoverer) that spent six years surveying Jupiter and its moons beginning in late 1995. With these missions, the solar system had begun to teach us an important lesson about diversity: the variety of possible worlds was far greater than we had dreamed.

Perhaps the biggest surprise was Io, a world slightly larger than our Moon and that orbits Jupiter just a little further away than our own satellite does from us. Despite this similarity in size and separation, the massively greater gravity of Jupiter compared to Earth has produced a very different history for these two worlds. The immense tidal forces produced by Jupiter bend and twist Io's body during its two-day orbit and create intense heating that makes Io the most volcanically active body in the solar system. In contrast, our Moon has been cold and geologically dead for billions of years. The most obvious result of Io's volcanic activity is its vivid red and yellow sulphurous colours. Although these are a little exaggerated in typical probe images, Io looks like a moon that Vincent van Gogh might have painted.

Europa is the next moon encountered as we move away from Jupiter, and it is potentially even more interesting because it is one of the most promising locations for life in our solar system. An ocean of liquid water lies beneath the moon's thin icy crust and this ocean is underlain in turn by a warm rocky sea floor. Europa therefore has the heat, liquid water and minerals that may be all that are needed for life to begin and thrive (sea floor volcanic vents are currently favoured by many experts as the place where life began on our own planet). Many worlds in the outer solar system contain water, but at these distances from the Sun it usually freezes harder than concrete. Tidal heating on Europa, similar to that on Io but less intense, has prevented this by melting a 100 kilometre-thick layer of water that lies a few kilometres beneath the visible surface. This is an underground sea with twice the volume of Earth's oceans and one that Jules Verne would have been proud to imagine. The evidence for an internal ocean on Europa comes from measurements of how this electrically-conducting, salty sea disturbs Jupiter's magnetic field. This interpretation was confirmed through observations of the Galileo probe's deflection by Europa's gravity as it swung by the satellite, which told us that the outer layers of Europa are not heavy enough to be made of rock. The best evidence of all is that the moon's appearance strongly supports the existence of a sub-surface sea; Europa's exterior looks like a partially melted and re-solidified ice pack. The huge, fragmented floes of this moon seem to have been shattered by immense pulses of heat from below or impacts from above and then rapidly refrozen into immobility by the bitter cold of space. The entire moon is also criss-crossed by hundreds of long, brown cracks stained by minerals carried in currents of warm water rising from the bottom of the sea. The best description I have heard of Europa is that, as a result of those coloured cracks, it looks like a huge ball of string.

Moving further out from Jupiter to the next moon, Ganymede, brings us to the largest satellite in the solar system; a moon bigger than the planet Mercury. At first sight, parts of this world look like my idea of a proper moon with a heavily cratered surface. But that impression doesn't last long; only about 40 per cent of the surface has that appearance. The rest of Ganymede's surface is a jigsaw puzzle of innumerable plates, each like a massively magnified fragment of a muddy, tyre-track-covered building site. These grooved plates have fewer craters than the un-grooved regions, implying that they are significantly younger, and so it seems that more than half the surface of this moon has been stretched and deformed by huge forces to produce a world that, like the Earth with its oceans and continents, is schizophrenically divided into two very different terrains. At present we have no firm idea about what exactly happened, when it happened or why it happened, but there is little doubt that this moon has experienced intense surface activity at some point in its history. Ganymede probably also has a sub-surface sea like that of Europa, but, since Ganymede is a significantly larger moon with a thicker water layer, the higher pressures at the ocean's base turn water into an exotic form of ice not found naturally on Earth (and hard to produce even in a laboratory). The liquid part of Ganymede's ocean is therefore sandwiched between normal ice at the moon's surface and this weird, high-pressure ice at the sea floor, so that it is not in contact with rocks rich in the mineral ingredients of life. This makes the sea of Ganymede a much less promising site for alien biology than Europa's ocean.

The outermost Galilean moon, Callisto, finally brings us to a world that really is dead, dusty and pocked with craters, but this moon-like appearance is deceptive and disappears on closer inspection. Callisto has the oldest surface in the solar system, since even our own Moon has been active more recently. The lunar maria, the dark features that give the appearance of a 'man in the Moon', are the result of flooding of huge impact basins by basaltic lava long after the craters were created during the violent collisions of the young solar system. In contrast, Callisto shows no signs of any geological activity at all following the intense bombardment of the solar system's half-billion-year birth pains. Callisto does have one intriguing mystery to offer: a surprising lack of small craters. The heavily cratered surfaces of the solar system usually have many more small craters than large ones and Callisto shows this pattern until you get down to craters about 3 kilometres across. Below this threshold, there are far fewer craters than you'd expect. So Callisto may, after all, have some kind of geological activity on its surface; some exotic process that obliterates smaller features. Alternatively, perhaps the young Callisto had a thick atmosphere that, like the Earth's, prevented smaller meteorites from getting through to the surface.

The unpredicted variety of the Galilean moons reflects, in miniature, the diversity we see throughout the solar system and beyond. All planets are peculiar in one way or another. In our solar system Mercury is made mostly from iron, Venus may have selenium snow, Earth has continents that move, Mars has huge volcanoes, Jupiter is much bigger than all the other planets put together, Saturn has its stunning rings, Uranus is tilted on its side, and Neptune is the true 'blue planet' of the solar system. No two worlds are the same, even when they have similar dimensions and share a similar location. Venus and Earth, like the moons of Jupiter, are close in size, started life with almost identical compositions and are near neighbours in space. Despite this, it would be hard to find a less pleasant and less Earth-like place than the surface of Venus, a planet that has justifiably been called 'our evil twin sister'. Even Uranus and Neptune, perhaps the closest our solar system comes to a true set of twins, are quite distinct when looked at in detail, with Neptune having a stormier surface and a more uniform internal construction than Uranus.

The unique character revealed by our robot probes of every world they have explored has been emphasised by the discovery in recent decades of exoplanets. The first widely accepted planet orbiting a normal star other than our own Sun was discovered less than twenty years ago by Michel Mayor and Didier Queloz. These Swiss-based astronomers detected the very slight changes in starlight caused by a planet orbiting the star 51-Peg, 45 light years from Earth. On a dark winter's night, 51-Peg, which sits half way down the west side of the square of Pegasus, is just visible to the naked eye, but nothing would make it stand out to a casual observer. More detailed study has shown that this star is remarkably similar to the Sun although it is slightly heavier, slightly brighter and probably just a little younger than our own star. More remarkably still, Mayor and Queloz were able to show convincingly that a planet half the size of Jupiter revolves around 51-Peg every four days in an orbit so tight that temperatures on the planet's sun-scorched surface must be close to 2,000°C. For comparison, Jupiter takes twelve years to orbit our Sun, and even Mercury, the closest planet to our star, takes nearly three months to complete each orbit.

The discovery of this, and subsequently many other, 'hot Jupiters' was a shock to planet-formation theorists. They thought they understood how the solar system formed and expected the same processes to occur around other stars. Small rocky worlds should form close to a star, with larger planets forming further out since, according to our theories, the near-star environment is simply too hot for gas to condense and make gas giants like Jupiter or Saturn. The now widely accepted explanation is that gas giants do indeed only form further out but that they can migrate inwards once formed. So, thanks to Mayor and Queloz, we now know that planets exist outside of our own solar system and we also know that other planetary systems can be very different from our own.

Detecting a planet, as Mayor and Queloz did, by looking for fluctuations in starlight is like trying to check on the breathing of a patient by listening to the siren of his departing ambulance. This may seem a rather whimsical way of putting it, but it's actually quite an accurate analogy. The drop in siren pitch as an ambulance rushes past is a familiar experience that results because successive waves of siren-sound crowd together to give a high pitch as an ambulance approaches but are stretched into a deeper tone by a receding ambulance as each successive pulse begins its journey to your ears from further and further away. The strength of this 'Doppler effect' depends on the speed of the ambulance and this could, in principle, allow a patient's vital signs to be detected from the sound of the siren alone. The passenger's breathing will rock the ambulance very gently and the resulting tiny fluctuations in speed minutely change the siren's pitch. However, the effect would be unimaginably minuscule, and swamped by the shaking of the ambulance as it trundled over the rough road, so that you'd have great difficulty convincing anyone that you had really detected a breathing patient.

In an almost identical fashion, the colour of light from a star is changed by the star's motion. The pulses of a light wave are crowded together, if the star is coming towards us, and this gives a tiny blue-shift in its colour. In contrast, there is a slight red-shift if it happens to be moving away. Any planets orbiting such a star will gently rock it back and forth to produce planet-revealing fluctuations in these colour-shifts. As with my ambulance analogy, the effect is tiny and swamped by noise, but very heavy planets with small, rapid orbits will generate the greatest rocking, and this technique therefore tends to find big planets in small orbits: hot Jupiters like the one orbiting 51-Peg. This approach to finding worlds beyond our solar system is called the radial velocity (RV) technique, and though there are other planet-hunting techniques that I'll describe later in this chapter, RV remains the most successful method of detecting exoplanets.

It's probably worth saying a few words at this point about planet-naming conventions. The world discovered by Mayor and Queloz is known as 51-Peg b rather than, as you might expect, 51-Peg a. This results from a long-standing convention for naming multiple stars, stars that come in pairs or bigger groups and orbit around each other. The nearest such star system to the Earth is Alpha Centauri, consisting of the Sun-like star Alpha Centauri A and the slightly dimmer Alpha Centauri B, which take 80 years to orbit around each other. There is a third, very faint, member of this system that should be called Alpha Centauri C but more often goes by the name Proxima Centauri since it happens to be the closest star to Earth. Planets are named using this same convention except that lower-case letters are used to indicate that the companion is a planet rather than a star. Hence 51-Peg b because 51-Peg A is the star itself. This naming convention can produce some real mouthfuls, with my favourite, so far, being OGLE-2006-BLG-109L b. I'll come back to this planet, and its companion OGLE-2006-BLG-109L c, at the end of this book when I try to explain this rather curious and complex name as well as why these particular planets are important.

Many of the newly discovered worlds will probably be given more euphonious names eventually but, for now, the vast majority follow this rather dull convention simply because there are an awful lot of them and there's been no time to agree on anything more poetic. We now know of hundreds of exoplanets (942 as I write this in September 2013). Five hundred and thirty-two of these have been found using the RV technique and, despite the observational bias of RV towards hot Jupiters, the star–planet separations range from worlds so close to their star that they orbit in not much more than a day, through to planets as far from their star as Neptune is from ours. The biggest two planets yet found are at least 30 times heavier than Jupiter and both orbit the star Nu Ophiuchus – which is itself a bit of a heavy-weight, being about three times more massive than our own Sun. These worlds are in fact large enough to be considered stars in their own right, brown dwarfs as they are known, and serve to show that planets' sizes span the full range right up to that of small stars.

In contrast, the smallest planets discovered so far are Earth-mass worlds such as one that orbits Alpha Centauri B. Sadly, this planet is 25 times closer to its slightly smaller sun than we are to ours and is therefore far too warm to be habit-able. If you could stand on this blisteringly hot world, a sun looking twenty times bigger than our own would dominate the sky, and Alpha Centauri A would appear as a second, dis-tant sun with a disc just large enough to be visible by eye. Proxima Centauri, on the other hand, is so faint and so far from Alpha Centauri Bb that it would not stand out in that planet's night sky, although it would be just barely visible if you knew where to look. Our own Sun would appear in the constellation Cassiopeia as one of the brightest stars in the, otherwise scarcely different, heavens of this near neighbour in space.

The RV method is not the only way in which planets have been found outside our solar system. The first exoplanets were found, a few years earlier than 51-Peg b, by detecting fluctua-tions in the arrival time of the rapidly pulsating radio signals from a pulsar, a dead star with the mass of a sun but the size of a city. Pulsars are created during supernova explosions, and any planets orbiting stars prior to these cataclysms should have been utterly destroyed. The pulsar planets must therefore have been formed from the debris and, as so often in this story, their existence was a complete surprise unpredicted by anyone before their discovery.

Another technique used to find exoplanets is to look for transits: regular dips in stellar brightness caused when a planet crosses in front of a star and blocks out a small fraction of its light. The effect is small and, for any given planet, unlikely to happen at all since the necessary alignment is quite rare. The chances of an Earth-like world producing a transit each time it orbits its star are about 1 in 700 and the drop in brightness it would produce is less than one-hundredth of a per cent but, nevertheless, Earth-like worlds can be found provided enough stars are looked at with sensitive enough equipment. So far, 333 planets have been detected by transits and there are thousands of candidate worlds awaiting confirmation by repeat observations. Most of these have been found by the Kepler space mission, a space-based telescope that has been continuously monitoring 100,000 stars in the constellation Cygnus since 2009. None of Kepler's discoveries are yet truly Earth-like but it is probably only a matter of time before it finds an Earth-sized planet circling within the habitable zone of a stellar system. Indeed, announcement of such a world could well come between me finishing this book and its publication (an occupational hazard of discussing topical subjects). Kepler has already found some pretty odd worlds, though, and perhaps the strangest are four circumbinary planets: planets that literally have two suns (note that Alpha Centauri Bb discussed above is orbiting only one of the stars in that triple-star system and therefore has only one sun).

Exoplanets can also be found by directly looking for them, but the enormous brightness difference between a star and any planet orbiting it makes this very difficult, with only 38 worlds having been seen this way by late 2013. However, unlike the RV and transit methods, directly looking for them works best for planets with large separations from their host star. The record separation, so far, is a planet with the mass of twenty Jupiters orbiting 500 billion kilometres from a star seven times the size of ours. Even allowing for its much brighter sun, this must be a very cold world indeed.

One final method for finding exoplanets is called microlensing. This technique detects changes in the brightness of distant stars caused when the gravity of a much nearer star focuses their light onto the Earth. The smooth increase and decrease in brightness, over a few weeks, that this chance alignment causes is distorted when planets are present around the nearer star. Analysis of this distortion allows the size and separation of the planet to be calculated. Microlensing has been used to find just 24 planets so far, but despite this low success rate, the importance of this approach will become clear later in this book.

Of all the techniques for finding worlds orbiting other stars, the transit method has the most potential to provide ground-breaking results in the coming decades. Transits are already providing the best data about the relative frequencies of different sizes of planets and different sizes of orbits. But transiting planets allow an even more thrilling possibility than just exoplanet discovery: the technique allows us to determine the composition of a planet's atmosphere even though we are hundreds of light years away. During a transit, light from the star passes through any atmosphere that the transiting planet possesses, and this allows its composition to be assessed. The drop in light during the transit is larger for those wavelengths of light where the atmosphere is opaque, and the wavelengths where any such absorption occurs can tell us about the make-up of the atmosphere. Such studies are in their infancy, with only hot Jupiters currently giving a strong enough signal to allow analysis. Nevertheless, two dozen worlds have had their atmospheres investigated and the constituents found include water, methane, sodium, carbon dioxide and carbon monoxide. This is currently the most exciting research area in exoplanet science and fascinating results are expected in the near future, especially once the James Webb Space Telescope (the Hubble Space Telescope's replacement) is launched in, or after, 2018.

Robotic exploration of the solar system, together with our recent discoveries of planets around other stars, has revolutionised our understanding of the variety of possible worlds. Until recently we divided planets into only two groups, the terrestrial planets (Mercury, Venus, Earth and Mars) and the gas giants (Jupiter, Saturn, Uranus and Neptune). Exploration of the solar system has shown us that this is inadequate and so we generally split the giants now into two groups: the gas giants Jupiter and Saturn and the misleadingly named ice giants Uranus and Neptune. This new category, ice giants, recognises that these worlds have significantly more volatiles (molecules such as water, carbon dioxide and nitrogen) than the gas giants. These volatiles are loosely described as 'ices' since they would have originally condensed as ice at these distances from the Sun, although, once incorporated into worlds like Neptune, they are certainly no longer in ice form. Exoplanet discoveries have also forced us to introduce additional planet categories to cover kinds of world not seen in our own solar system such as the hot Jupiters already discussed and super-Earths (rocky planets with diameters approximately twice that of Earth).

Nevertheless, these additional categories of planets still don't encompass the variety seen when worlds that are not strictly planets are also included. As we saw above, the moons of Jupiter show entirely unexpected ways of making worlds, and I haven't even mentioned the satellites of Saturn such as Titan with its ethane rain, rivers and lakes, Enceladus with its powerful geysers blasting water into space, or Iapetus with its 'Great Wall', a unique ridge running right around its equator. I also haven't mentioned the new category of 'dwarf planets' which had to be reluctantly introduced when it became clear that there are probably hundreds of Pluto-sized objects orbiting in the distant parts of our solar system. Dwarf planets are objects big enough for their gravity to make them spherical but not big enough to have cleared other large bodies from the surrounding space. Thus

Ceres, discovered in 1801 as the first of the asteroids that orbit between Mars and Jupiter, is now classified as a dwarf planet, as is Eris, which was discovered in 2005 orbiting our Sun three times further away than the slightly smaller Pluto. Pluto remains a fascinating body regardless of what we artificially choose to call it, and NASA's forthcoming New Horizons space mission can only further emphasise the restrictiveness of current planet classification schemes when it gets to Pluto and its five (or more!) moons in 2015. I can't wait to see the pictures.

The study of planets, moons, dwarf planets and exoplanets has revealed an unexpected variety of worlds. Fortunately, order can be brought to this complicated picture. From what we have seen in the solar system, the developmental history of any world largely depends on its size, its composition and its energy sources. Is a planet metal-rich, rocky, gas-rich or volatile-rich? Is it bathed in warm sunlight or does it shiver at the frigid edges of its system? Does it have significant internal heat from radioactivity and tidal forces or primeval heat left over from its cataclysmic formation? Taking six components (metal-mass, rock-mass, volatile-mass, gas-mass, illumination and internal heat), and crudely splitting worlds into small, medium or large in each of these categories, gives 216 different possible types of world. Of course, some combinations will be more common than others but our experience with the variety seen in the solar system and beyond suggests that examples of many of these categories could be found eventually. It seems to me almost inevitable therefore that we will eventually end up with hundreds of types of world as our knowledge of their variety continues to expand. In itself, the existence of many types of worlds does not necessarily mean that life-friendly ones are rare, since several different categories may be suitable, such as Earth-like worlds and Europa-like worlds. On the other hand, if small differences are sometimes important, the crude division discussed above may not capture the full richness of planetary evolution.

Venus, for example, is an Earth-like world in terms of its size, composition and location but it is certainly not life-friendly. The relatively small difference in solar heating may explain this, but other factors also play a role. In particular the Earth's magnetic field, which is uniquely strong and long-lived among rocky worlds in the solar system, could be important. Planets like the Earth that retain a strong field over billions of years may be more conducive to life because a magnetic field acts as a shield against the solar wind. The solar wind is a million-mile-an-hour stream of energetic particles sent out by the Sun that, if not deflected, can remove water from a planet's atmosphere as collisions between the wind particles and water split the molecules into their constituent hydrogen and oxygen atoms, allowing the lightweight hydrogen to escape into space. This probably happened to Venus, which has a thick but completely dry atmosphere and no magnetic field. Unlike Venus, our planet has a geomagnetic field that has persisted for billions of years. The reasons for this are still being investigated. It may have to do with disparities in rotation rate, composition or size. Whatever the cause, the existence of our magnetic field illustrates that there are differences in planetary characteristics that are not captured by the crude scheme outlined above but that may still be important to habitability.

Habitability may also be affected by properties not directly related to planet type. For example, location in the galaxy may matter (are there lots of nearby supernovas?), and the lifetime of a star is almost certainly a major consideration (will the planet be habitable for very long?). Furthermore, this book concentrates in particular on the influence of climate and the many factors that affect it, such as a planet's rate of rotation, the circularity of its orbit, the angle between a planet's axis and its orbit and, as we'll see in detail later, even the locations of other planets in the system. This gives another six properties to add to the six I mentioned earlier and, if planets need even moderate

fine-tuning of these to be highly habitable, a low probability
that any given planet will be life-friendly becomes mathematically inevitable.

Imagine taking all the planets that satisfy one of the properties – say, having the right amount of rock – and, for the sake of
argument, imagine that one in ten planets satisfies this criterion.
How many of these will also have the right amount of volatiles?
Again, for sake of argument, imagine that one in ten worlds
already chosen satisfies the volatile criterion too. So, one world
in a hundred satisfies both properties. If this argument is carried
through all twelve properties then, assuming that at each cut
we keep one in ten planets, we end up with only one planet in
a trillion satisfying all twelve properties. I should emphasise that
the numbers I've used here are for illustration only. I do not
know how many properties of the Earth are fine-tuned for life
and I do not know what is the probability of each of these happening by chance. So, please don't take my 'one in a trillion' too
seriously; the true frequency of habitable worlds could be much
higher than this or it could be very much lower. Nevertheless,
unless there are surprisingly few planetary properties necessary
for complex life, habitable worlds are going to be pretty rare.

The flip side of this argument is that there must be huge
numbers of planets in the Universe to ensure that, despite poor
odds, worlds like ours with the right combination of characteristics will still occasionally appear. So how many planets are there
in the Universe? We don't yet know if the majority of stars have
planets orbiting them but it is at least clear that planets are not
at all unusual. We also know that some stars have more than
one planet. Hence, a first guess is that there are roughly as many
planets as there are stars in the Universe. But how many is that?

Our galaxy is a disc about 100,000 light years across and
several thousand light years thick, giving it a volume of about
10 trillion cubic light years. In the vicinity of the Sun, there
is about one star in every hundred cubic light years. A crude

calculation therefore suggests that our fairly typical galaxy holds 100 billion stars. More detailed calculations suggest there are actually about 200 billion stars making up our galaxy. Furthermore, huge numbers of galaxies can be seen through our telescopes. An idea of just how many galaxies there are can be gauged from an extraordinary picture taken by a unique telescope during the Christmas holidays of 1995. Five years after the Hubble Space Telescope was launched, it was pointed continuously at a small patch of sky for ten days just to see what was there. A particularly boring spot was deliberately chosen in the constellation of Ursa Major (Latin for the Great Bear but perhaps better known as the Big Dipper or the Plough). The location needed to be dull; the Hubble Deep Field, as the picture would be called, had to look into the distant Universe without having its view obscured by relatively nearby objects. Over that ten-day period the telescope took several hundred digital pictures, which were added together by computer to produce the final, highly sensitive image. The result was stunning: 10,000 galaxies in a field of view so small that it would take a hundred of them to cover a patch of sky the size of the Moon. So next time you look at the Moon, remember Neil Armstrong but also remember that, wherever it is in the sky, there will be about a million galaxies hiding behind it! How many Moon-sized patches of sky do you think it would take to cover the entire heavens? Hundreds? Thousands? Actually it's about 20,000 and so, if we got Hubble to survey the entire sky, it would see tens of billions of galaxies. More detailed versions of this calculation imply that there are around 100 billion galaxies in the observable universe.

Hundreds of billions of galaxies each containing hundreds of billions of stars implies more than 10,000 billion billion (10,000,000,000,000,000,000,000) stars in the visible Universe, and the number of planets is probably similar. Numbers this large are notoriously difficult to visualise but, to get some idea, imagine sand so fine that you can barely see the individual grains. A pint glass would hold nearly a billion of them; a number so huge you couldn't pull them out one at a time even if you were insane enough to dedicate a sleepless lifetime to the job. Now imagine having enough of these sand grains to fill a box a mile long by a mile wide by a mile high. That'll be about the right number.

Even that doesn't complete the tally for the number of stars in the heavens. If currently favoured models of cosmology are correct, the part of the Universe that happens to be visible from the Earth, the observable Universe, is an insignificant fraction of the entire Universe. The bit we can see goes out 13 billion light years. More accurately, the furthest visible galaxies were 13 billion light years away from us when the light we now see started its journey, but they're now more than 20 billion light years away. However, if the latest cosmological theories are right, there are stars and galaxies much further away than this; but there has not yet been enough time, since the creation of the Universe less than 14 billion years ago, for their light to reach us. So, the whole Universe is larger than the observable Universe and it could be much larger. These new cosmological theories, which are very good at explaining many important features of the bit of the Universe that we can see, suggest that the whole Universe is truly vast and extends over scales that make the visible Universe look microscopic by comparison. In such theories the volume of the entire cosmos is at least a billion times larger than the bit we can see – and that's just the minimum estimate!

So, taking 100 billion stars per galaxy and 100 billion galaxies in a visible Universe that fills only one billionth of the volume of the whole Universe gives a minimum estimate of 10,000,000,000,000,000,000,000,000,000,000,000 stars and planets. To visualise this truly staggering number you will need to imagine enough fine sand to make an object the size of the

Moon. The number of planets in the whole Universe therefore makes even the number of grains in a cubic mile seem like small change, and this unimaginably large quantity of planets allows highly habitable worlds like ours to be both very rare and, at the same time, very numerous. As I said before, this is not a contradiction. One such world on average in a volume the size of the observable Universe is pretty sparse but still allows at least a billion of them in the entire Universe. Even if life-friendly planets are vanishingly rare, the Universe is so enormous that they remain inevitable. Under such circumstances we cannot draw conclusions about habitable worlds flooding the galaxy just because we happen to be sitting on one.

It's time now to take our eyes off the heavens to look at something much stranger: Earth itself.

4

Constant Change

Don't look up, look down! To understand life on other planets you need to look down at the rocks as well as up at the stars. Reading rocks to learn about alien life may seem about as sensible as reading tea leaves, but the history of our planet is written into its crust and that history tells us much about the conditions necessary for life. This history also tells us that there are many ways of courting disaster but that – extraordinarily – the Earth has avoided them all. In particular the Earth has followed a narrow path that has avoided climatic catastrophe. Unlike our immediate neighbours in space, the Earth has not become too hot, like Venus, or too cold, like Mars.

Is 4 billion years of good weather on Earth really so surprising? Well, yes it is. The Earth is incredibly old and so there has been ample time for even the slowest of changes to produce catastrophe. A warming trend as small as 1°C every 100 million years would have been enough to make our world uninhabitable by now, and it would not have been surprising had such a trend occurred. The oceans, atmosphere and land surface of our planet have changed continuously and dramatically throughout Earth's history, as has the amount of heat emitted by our Sun. The details of exactly how much change has occurred, and when, are the subject of intense scientific debate but no one doubts that there have been profound alterations to the Earth's environment. Any of these changes could have caused devastation but, instead, they usually managed to roughly cancel each other out.

Before I describe these changes in a bit more detail, I want to say a little more about the immense lengths of time involved.

Unfortunately, it is practically impossible for the human mind to really grasp the scale of deep time. Nevertheless, many writers of geological and astronomical books have tried. A frequent approach is to use distance to represent time. For example, if every year is replaced by a millimetre, the age of the Earth is equivalent to the distance from London to New York. Personally, I don't find this at all helpful because I have never walked from London to New York and have no real feeling for how far it is. Another approach is to shrink time by, for example, imagining that one second represents one year. On this scale, I was born 53 seconds ago, the pyramids were built about an hour before that and the dinosaurs died out two years earlier. Using this reduced timescale, the Earth formed about 150 years ago. Shrinking each year down to a second works reasonably well for me but I still find it very abstract. Sadly, doing better is an almost impossible task.

The greatest length of time for which anyone can really claim a concrete grasp is probably a human lifetime. The oldest man I ever met was my great-grandfather who was 85 years older than me. If, when he was a small boy, my great-grandfather had also known someone 85 years older, that person would have been born in 1790. I am therefore separated from the French Revolution by only two human lifetimes, which makes that particular piece of ancient history seem quite recent. Using an 85-year human lifetime as my yardstick, I am separated from early feudalism by ten lifetimes and the beginnings of civilisation by a hundred lifetimes. A few thousand lifetimes link me to the earliest modern humans. At this point, I feel that the human-lifetime yardstick has started to become useless. To go back a truly geological length of time, say to the age of dinosaurs, we need a million lifetimes. I find it impossible to relate to such an enormous stretch of time but the extinction of the dinosaurs is, in turn, a relatively recent event. More than 98 per cent of Earth's history occurred before they disappeared.

So, I'm afraid I'm going to have to give up after all, but perhaps that's the point. The lengths of time we are looking at in this book are so far beyond human imagination and experience that almost anything could have happened. The time periods were long enough for hundreds of mountain ranges to grow and be ground back to nothing; and it was long enough for the descendants of bacteria to change so much that they could build rockets that travel to Mars. To do justice to all the major events in the life of our planet would require an entire library; I'll outline just a few of the more important ones. The key point to bear in mind throughout this chapter is that once life appeared, and despite the enormous changes the Earth went through, the average surface temperature did not vary by more than a few tens of degrees centigrade.

Let's start around 4.6 billion years ago when there was no Earth, no Sun, and no solar system. Instead, there was just a slowly spinning cloud of interstellar gas and dust, spread over dozens of light years, a nebula large enough to build hundreds of stars and their accompanying planetary systems. Slowly, over millions of years, thicker parts of the cloud became even denser as their slightly higher gravity pulled gas and dust in from their surroundings. The cloud had started to collapse. As material converged on each of the more compact regions, it began to spin rapidly around in much the same way that water spins as it converges on the plug-hole of an emptying bath. As in a bath-tub whirlpool, this rapid spin pushed material away from the centre of convergence and collapse all but ceased. What happened next is still poorly understood, although we see the process going on today in places like the Orion nebula a few hundred light years from the Earth. Spin was somehow transferred out of the central clumps and into orbiting discs of gas and dust, looking like the rings of Saturn but a million times larger. The reduced rotation in the inner regions then allowed these cores to shrink further and soon they became small,

dense and hot. Within a few million years, central temperatures exceeded 7 million degrees centigrade, allowing nuclear
reactions to begin. One by one the members of a brand-new
star cluster ignited.

Our Sun was one star in that cluster. It would have been
inconspicuous and unspectacular compared to the most flam-
boyant of its siblings but it was, in fact, significantly larger
and brighter than average. The very brightest members of the
cluster burned themselves out within a few million years. In
contrast, our more frugal star was destined to shine for 10 billion years. In fact, the young Sun was even more frugal then
than it is today. As the Sun has aged, the hydrogen it uses for
nuclear-fusion fuel has been consumed and an inactive core
of helium has slowly grown at its centre as this 'nuclear ash'
has sunk through the lighter hydrogen. Present-day hydrogen
fusion is therefore restricted to a zone surrounding an inactive core and, rather counter-intuitively, this has had the effect
of increasing the total energy produced as our star has grown
older. As a result, the young Sun radiated only about 70 per
cent of the heat that it does now. Despite this, the Sun's visible
surface has not actually become much hotter. Instead, like a
slowly inflating balloon, it has expanded to become a bigger
radiator able to pump out more heat. If you could have looked
at our Sun 4.5 billion years ago you might have noticed that it
was a little redder but you would certainly have noticed that it
looked smaller.

Looking at our juvenile star you would also have been
struck by the beautiful disc of gas and dust that still surrounded
it, but this beauty was short-lived. Clumping began again, on a
planetary rather than stellar scale, as rocky and icy bodies in the
disc collided and stuck together. Gradually, the larger bodies
consumed the smaller ones and, within a few million years, collisions involved bodies the size of moons and planets. A particularly violent collision nearly demolished one planet as it smashed

into another, slightly smaller, one. Most of the debris from this impact was flung into distant space, but some remained to form a ring of rubble orbiting 10,000 kilometres above the surface of a world that had only barely escaped complete destruction. That traumatised planet was the young Earth and the rubble-ring condensed in less than a thousand years to become our Moon. The Earth–Moon system was therefore born in great violence and, as we shall see in detail in a later chapter, the subsequent evolution of this system has had profound effects on the history of our planet.

The collision that formed the Moon was not the last impact our planet suffered, although it was quite possibly the largest. Intense bombardment of the Earth continued at a gradually declining rate for another 500 million years. Even today the Earth suffers collisions with bodies several kilometres across every few million years or so. However, in those early days impacts of that size occurred many times a day and truly enormous collisions must have happened every few years. These collisions generated a great deal of heat and, for 200 million years, the Earth's surface was too hot for liquid water and the atmosphere mostly consisted of steam. Then the rate of collisions dropped, the surface temperature cooled and atmospheric water started to condense. A 100 million-year rainstorm began and oceans flooded our planet for the first time.

Life appeared within a few hundred million years of liquid water. We do not know exactly when, where or how life emerged on our world but we know that it happened early on because we have convincing fossils from 3.5 billion years ago and fairly convincing chemical signatures of life in rocks approaching 4 billion years old. As for how it happened, one speculation is that increasingly complex chemical reactions in the tiny spaces between hot, rocky grains buried beneath the volcanically active sea floor led to the emergence of a molecule that could make copies of itself from the chemicals in its surroundings. The first self-replicator had therefore appeared. This very first replicator made a copy of itself and so there were now two. But these two also copied themselves to give four and these reproduced to give eight and so on until, in a very short time, the number of replicators was enormous. A complex series of chemical reactions that had taken millions of years to happen once, by chance, quickly became widespread. This was not yet life as we know it, but it was a start.

There are many variations to this story. Charles Darwin speculated that life may have emerged in some warm, muddy pool. Others have suggested that it happened in shallow seas or even in the clay of river beds. Perhaps most speculative are proposals that life didn't form on Earth at all but was imported from elsewhere, Mars for example. Whatever the details, the key step was the appearance of entities that make copies of themselves. Self-replicators were the beginning.

However they emerged, the first primitive self-replicators were not very good copiers and many mistakes were made. Most of the badly copied replicators were not viable and their lines died out immediately but some of the mistakes were not fatal and so, after a short while, many varieties of replicator existed. Some of these different species of replicator just happened to be better at scavenging their chemical constituents from the dilute sources available and they thrived while their less efficient cousins died out. In other words, natural selection had begun. As a result the replicators rapidly became very efficient and accurate makers of replicators. Mistakes were still made sometimes, and occasionally these mutants proved to be even better than their parents so that a new variety took over the world, for a while. At some point, replicators emerged that were able to steal resources from what had become the widest available source, their fellow replicators. These first predators thrived until all the easy prey were gone and only the fortuitously less-digestible replicators were left.

A great number of chemical and physical innovations then occurred in succession since any advantage, however slight, produced a population explosion for the innovator. Within a very short time all replicators were stored within tiny fatty bubbles that concentrated life-giving chemicals and made the replicators safer from attack by other replicators. Single-celled microbes had appeared. By 3.8 billion years ago sophisticated bacteria infected the crust, land surfaces and oceans across the entire planet. Then the first of several disasters struck the nascent biosphere.

A period of renewed heavy bombardment began as large objects were deflected into the inner solar system by recently formed outer planets. The evidence for this is visible on any clear moonlit night. The large, dark marks that make up the features of 'the man in the Moon' are easily visible to the unaided eye. Our ancestors also saw these features and believed them to be seas on the surface of the Moon. In reality, they are enormous impact craters – and analysis of the rocks, brought back by the Apollo astronauts, shows that these craters formed 3.8 billion years ago. It seems obvious that similar-sized impacts must have also happened on Earth and that these would have sterilised the surface so that life only just clung on in the deep, dark places of the world. Much of the atmosphere and ocean would have been blasted off our planet in these cataclysms, to be slowly replaced, over millions of years, by exhaust from volcanoes.

Despite this catastrophe, by 3.5 billion years ago the Earth was a vibrant yet still alien world. The Sun, slightly redder and noticeably smaller than today, shone through a cloudless but hazy sky onto a warm salt-water sea covering most of the planet. In many places the sea floor was lifted by massive flows of hot magma moving deep below the surface of the hot young Earth. The resulting shallow seas were carpeted by countless bacteria in colonies large enough to form reefs. Fierce tides ebbed and flowed across these microbial mats every seven hours because the days were short and the Moon close. Single-celled organisms floated in the planet-wide ocean, grew on the deep sea floor, or lived deep within the Earth's crust.

These early bacteria obtained their energy and their building materials from the chemicals around them but these were available in only a few places. Life's conquest of the rest of the planet had to wait until photosynthetic microbes evolved. This was followed by the appearance, around 2.8 billion years ago, of the first organisms to give off pure oxygen gas: cyanobacteria. It took hundreds of millions of years for the full consequences of this to develop, but over that time, oxygen started to become an important constituent of the atmosphere. The geological evidence for a slow, but eventually substantial, increase in atmospheric oxygen is robust. Sediments deposited more than 2 billion years ago frequently contain banded iron formations (BIFs) but they virtually disappear from the rock record after this time. BIFs cannot form in the presence of oxygen. However, it is not clear exactly how much oxygen there was in the atmosphere at various times in our planet's history. We are fairly sure that there was hardly any 2.2 billion years ago and that, by half a billion years ago, oxygen levels were close to modern values. But the experts continue to debate among themselves exactly how fast our atmosphere got from one state to the other.

Even if the rise in oxygen was near to the lowest estimates, oxygen-intolerant microbes, which included most life-forms at the time, soon became restricted to special environments where oxygen was absent. These anaerobic bacteria moved to where they still live today – in buried rotting matter, deep in the crust, or in ocean depths undisturbed by oxygen-carrying currents. Newly evolved oxygen-breathing bacteria took over most of the world. This drastically altered atmosphere had a profound effect on global temperatures and, as we'll see in later chapters, produced the most unstable climate our world has ever seen.

It was also about 2 billion years ago, give or take a few hundred million years, that a dramatically new life-form appeared. These new organisms were so radically different from their bacterial forebears that, by comparison, oak trees and people are close relatives. Criminal behaviour was at the root of this innovation. Many life-forms, including humans, prefer theft to hard work since it's easier to take food from another organism than it is to make it yourself. Large organisms do this by swallowing small ones and small organisms do it by infecting large ones. However, it's not always clear which organism is predator and which is prey. It is particularly unclear when, as was the case for all life at this time, the organisms are simple, single-celled ones. Is a small germ inside a larger one being eaten or is it a pathogen?

When this situation first occurred an arms-race ensued. Sometimes the predators had the advantage, sometimes the infectors. Eventually a mutually beneficial truce was arrived at – or, more accurately, some groups of organisms stumbled by chance upon an arrangement that immediately made them very successful and so they thrived and became numerous. Certain small microbes became welcome, permanent residents of larger ones. They then had access to plenty of nutrients and could avoid being swallowed by less friendly predators. In return, the small microbes provided chemical services for their hosts. These aggregated organisms had taken life to a whole new level of organisational complexity.

In the earliest of these symbiotic relationships, the small microbes were the oxygen-breathing bacteria I mentioned earlier, and so they gave their hosts the ability to use oxygen gas. The resultant compound creatures were the common ancestors of amoebas, mushrooms, oak trees and people. A little later some of these organisms also took on board cyanobacteria and gained the trick of photosynthesis. The resulting algae were the direct ancestors of all plants. The differences between the simpler, pre-existing life-forms and these newer organisms are so profound that biologists treat this division as the most fundamental one in the tree of life. Simpler organisms are called prokaryotes and the compound creatures are called eukaryotes. Plants, animals and fungi are merely different sorts of multi-celled eukaryotes and so, in this scheme, are relatively close cousins.

At roughly the same time as all this was going on, but possibly much earlier, the planet's interior underwent one of the most important transitions in its history: the appearance of plate tectonics. To understand plate tectonics, we need to look at one of the greatest scientific revolutions of the 20th century. This did not take place in a scientific field like theoretical physics or biology, which are rather prone to bouts of radicalism. It took place in the usually well behaved field of geology. After hundreds of years of scientific study we suddenly realised that a single, simple fact was responsible for virtually all the geological processes we could see on the surface of our planet. Geological activity almost all results from the fact that the continents slowly move across the surface of the Earth.

This book is full of coincidences that aren't coincidental, and the discovery of plate tectonics starts with one of these. Sixteenth-century European explorers were the first to notice that the eastern coastline of South America fits the western coastline of Africa like two pieces from a giant jigsaw puzzle. This coincidence is reinforced by the much more recent discovery that even the rock types are the same when we compare outcrops in South America to those in the equivalent parts of West Africa. It seems obvious that, at one time, the pieces must have been stuck together. For most of the 400 years since this was first suggested the match between the two coastlines was dismissed as a simple coincidence, although bizarre explanations were occasionally offered. Some crackpots, for example, suggested that the Earth used to be smaller and that Africa and

South America had been forced apart as the planet expanded. Other madmen were convinced that entire continents simply drifted across the surface of the Earth. Convincing evidence that the second group of lunatics were completely correct was finally obtained in the 1960s when we first began to explore the deep ocean floor.

The mechanism of continental drift is quite straightforward. New ocean floor is constantly being created at ridges that lie roughly along the centre-lines of each ocean. Actually, to call them ridges doesn't really do them justice, since they form a 70,000 kilometre-long mountain chain covering nearly a quarter of the Earth's surface. This mountain chain is, by far, the largest on Earth but we were unaware of it until recently because it nearly all lies a kilometre or more beneath the surface of the sea. Mid-ocean ridges form as molten material wells up from the hot interior of the planet and cools to form brand-new sea floor. This new ocean crust slowly moves away from the ridge centre to be replaced by even younger magma oozing up behind. The ocean floor therefore drifts away from its point of origin, at a speed of about a metre every few decades, for a few hundred million years until it gets to a point where it is so old, cold and heavy that it sinks back into the Earth to be re-absorbed by the molten interior. This sea floor destruction process is called subduction and it's the tug of the heavy ocean floor, as it sinks back into the Earth's interior, that largely drives plate tectonics.

Continents ride on the resulting system of moving oceanic plates and are therefore slowly transported around the surface of the Earth. However, the oceanic plates have not always had continents sitting on their backs. Rather, continental crust has been slowly created as a by-product of subduction. One way this happens is that volcanic residues deposited on the surface of oceanic plates pile up at the subduction zones, in a process called accretion. New Zealand, for example, is growing this way today as islands and submerged sea-mounts are scraped off the Pacific plate as it sinks beneath the east coast of Te Ika-a-Maui (North Island). However, the more important continent-forming process, arc volcanism, occurs because adding water to hot rocks makes them melt. Most minerals contain small amounts of water bound tightly into their crystal lattices and the effect of this is to lower the melting point substantially compared to a dehydrated rock with an otherwise similar composition. The rocks in a subducting plate contain more water than material within the Earth's interior and so they tend to melt very easily as the plate subducts. This produces magma, which rises to the surface to produce intense volcanism and, as a result, brand-new land. As with continent growth by accretion, continent growth by arc volcanism is still going on today. For example, the islands of Indonesia were formed, and continue to grow, by this mechanism. Subduction-related volcanoes are particularly explosive and dangerous, good examples being the volcanoes in the Andes, in the north-west of North America and in Japan.

The net result of both accretion and arc volcanism is that light volcanic residues slowly accumulate at the Earth's surface. The Earth's continents are entirely made from these residues along with the results of reworking of this material into sedimentary rocks. Continent construction has been going on for billions of years and the area of the continents has therefore slowly increased through time. Four billion years ago there was hardly any dry land at all; whereas now, one third of the Earth's surface is continental. The resulting continents are too light to sink into the Earth's interior and so, compared to the oceans, they are almost immortal. This lightness, together with the greater thickness of the continents, also means that the continental surfaces sit much higher than the ocean floors, typically 5 to 10 kilometres higher, and therefore stick out above the sea. To a geologist, the fact that oceans are wet and continents dry is a minor side-effect rather than their defining characteristics.

Sometimes a new ocean-spreading centre forms inside an existing continent. This is happening today along the Great Rift Valley of East Africa, which, at its northern end, joins up with a brand-new ocean called the Red Sea. Something very similar occurred 190 million years ago, in what is now the South Atlantic, resulting in a single giant continent splitting apart to form the two separate continents of South America and Africa. So, those 16th-century explorers were right – South America and Africa do fit together very well and this is not a coincidence at all. The slow movement of Earth's tectonic plates caused South America and Africa to separate in a process that continues today. From the point of view of this book, plate tectonics is important because it has caused a steady increase in the amount of dry land and, as we will see later, because it recycles chemicals between the Earth's interior, its surface and its atmosphere. Continental drift and the cycling of materials both have a profound effect upon our climate.

And so, by 2 billion years ago when our planet was about half its present age, Earth had developed a substantial amount of dry land, the Sun had reached 85 per cent of its modern heat output, and the Moon had drifted away from Earth to almost its present distance. The climate at this time was experiencing a period of great instability and our biosphere was undergoing the most profound changes since the appearance of life itself. The stage was set for the final major biological act in the creation of modern Earth, the appearance of complex multi-celled organisms.

The evolution of multi-cellularity happened not once but at least three times, with the independent evolution of multi-celled plants, multi-celled animals and multi-celled fungi. Indeed, some experts claim that multi-celled organisms appeared, independently, as many as ten times. Colonies of single-celled organisms had been common for most of life's history but a multi-celled organism is much more than just a colony of identical cells. The cells in a truly multi-cellular creature are differentiated: they take on different tasks while remaining genetically identical. The cells in my arm muscles are very different from the cells forming my brain but all these cells have a nucleus containing exactly the same genes. Although the ability to produce these different tissues took a long time, perhaps as much as a billion years, the eventual appearance of multi-celled organisms had a very profound effect on the entire planet. The evolution of large plants had the most significant consequences, because once the resulting organisms conquered dry land, they quickly covered much of the continental area, causing big changes in its heat-absorbing properties and also in the rate at which it was eroded. The appearance of these plants also allowed the eventual colonisation of the land by animals and, half a billion years later still, the emergence of human beings.

By the time land plants appeared, the Earth's story, so far, was almost 90 per cent complete, but life large enough to be visible had only just emerged. Most books on palaeontology only start to get into their stride just before this point. The emergence of animals large enough, and with body parts hard enough, to leave easily visible fossils resulted in the, geologically speaking, instantaneous appearance of animal life about 540 million years ago. The history of life since that time is familiar, with the emergence of fish (500 million years ago), amphibians (400 million years ago), reptiles (350 million years ago) and mammals (240 million years ago). Note that this story is completely biased towards our own line of descent. How many people know as much about the evolutionary history of trees or spiders? In any case, from the point that animals appeared, the story starts to get bogged down in petty details. Most of the really important evolutionary innovations occurred before even fish evolved. Quite frankly, our climate has not been strongly affected by the peculiarities of vertebrate evolution at all and so I'm not planning to say any more about it.

To conclude, I should in all honesty admit that experts would argue over almost every one of the details in the story I have just given, but all would agree that the Earth has seen dramatic transformations many times in the last 4 billion years. As we shall see in the next few chapters, these changes in land, sea and air could have altered the temperature by hundreds of degrees centigrade but, fortunately for us, this didn't happen. So, the story of our home planet is a story of constant and massive environmental change and, despite this, also one of reasonably stable temperatures.

5
Air Conditioning

The Earth's climate system is a complex one in which rocks, air, water and life interact to influence temperature. This chapter will begin our look at how that system works. I'll examine the factors that control how hot the world is on average and at how these factors can change through time. In particular, I'll take a look at the notorious 'greenhouse effect' since this is as important for understanding the long-term history of our climate as it is for understanding how the world's climate will evolve in the near future.

Burrator Reservoir, in south-west England's Dartmoor National Park, is a good place to start. My university department takes its new geology students to this spot every autumn to give them their first taste of intrusive volcanic rocks – rocks formed when molten magma flows through the Earth's cool upper crust slowly enough to solidify before it breaks through to the surface. The uplands of Dartmoor exist only because the resulting granite, deposited near the beginning of the Permian Period 290 million years ago, is more resistant to erosion than the softer rocks of the surrounding, low-lying countryside. Our students first see the granite in a small abandoned quarry, just south of Burrator Reservoir, and this location illustrates nicely many crucial components of the Earth's climate system. The geological processes operating in this area act like a thermostatically controlled air conditioning system and, together with similar processes occurring in many places across the world, help keep temperatures on our planet roughly constant and, hence, suitable for life.

The Burrator quarry dates back to 1923 when granite extracted here was used to raise the dam height of the nearby reservoir. The quarry used for the original dam construction, in the 1890s, had disappeared beneath the waters of the resulting reservoir and a new quarry had to be excavated when expansion of nearby Plymouth demanded a deeper reservoir. This new quarry was therefore used for only a few months and, today, it has become a small car park for visitors to the area. As you drive in, a footpath can be seen in the far left corner, climbing up the quarry face in front of you. Walk along this path for a minute or two and you will come to an exceptionally beautiful spot where a bench has been placed in memory of a previous admirer of the stunning views. If you get the chance to visit, sit on the bench and absorb the beauty and serenity of the area. Looking north you'll see the rippled surface of the Burrator Reservoir through gaps in the trees, with the craggy summits of Peek Hill, Sharpitor and Leather Tor dominating the horizon beyond. To the east another rock-strewn hill, Sheep's Tor, rises above a square church tower peeking through the tree-clad slopes surrounding the reservoir. The scenery changes dramatically to the south and west, however, where there are lower, more rounded hills bereft of rocky outcrops. If you're lucky enough to visit on a very still day, the sound of nearby water cascading down a stream can be heard as the reservoir outflow makes its way along the River Meavy to the sea fifteen miles south of the quarry.

The landscape around Burrator Reservoir is therefore one of water, trees and 'tors'. *Tor* is a Celtic word meaning hill and many of the peaks in western Britain retain this ancient name despite over a thousand years of Anglo-Saxon dominance. However, 'tor' has also come to denote the outcrops of granite that crown many of the hills in this part of the world. There is a small example of a granite tor right next to the bench at the viewing point above the quarry. This miniature tor consists of 30 or so rounded, metre-sized blocks of stone that seem to grow out of the hilltop. Tors like this formed over hundreds of thousands of years when the granite was still buried deep underground. The subterranean rock was gradually dissolved as mildly acidic rainwater trickled down from the surface through fractures in the granite, and when it eventually became exposed by the slow wearing-down of the hilltop, the corroded material around these infiltrated cracks crumbled away to leave fragmented blocks. The boulders on the craggy hills to the north and east have all formed in this way, but the boulder-free hills to the south and west are round and low because there is no granite in those directions.

Rain is able to dissolve the granite around Burrator, and other volcanic rocks elsewhere in the world, because it has absorbed carbon dioxide from the air and become slightly acidic as a result. I'm not sure that, as the song has it, 'for every drop of rain that falls, a flower grows'. In fact, I'm pretty sure that even the thriftiest of flowers uses more than a few cubic millimetres of water but, more prosaically, I do believe that for every drop of rain that falls a little carbon dioxide is removed from the atmosphere. And carbon dioxide is, of course, a greenhouse gas.

The greenhouse effect plays a central role in the Earth's climate system. One way to understand the greenhouse effect is to imagine looking at the Earth from space using infra-red goggles. From this infra-red view the planet looks like it is glowing. You can see the Sun glow without such goggles because it is literally white-hot – so hot that it glows in visible light. Slightly cooler objects such as barbecue coals are red-hot, which simply means that that they are not hot enough to emit much yellow or blue light. Cooler objects than hot coals fail to emit even red light but, with infra-red goggles, you can see the more-red-than-red light that they do emit. Viewing Earth from outer space while wearing infra-red goggles might sound rather interesting. Actually, it isn't remotely interesting. In infra-red light the world is a uniform, featureless, haze-shrouded globe because the lower atmosphere is opaque to infra-red radiation and you can't see through the fog to the surface where all the interesting features like mountains, seas and people are. Infra-red radiation from the warm surface of the Earth is captured by this low-altitude haze and does not easily escape into space. This trapping of infra-red near the surface causes warming at ground level – a concept that has been understood by scientists for a century and a half. The man who did most to make this clear was John Tyndall, an Irishman born in 1820 who became superintendent at the Royal Institution, the prestigious London-based organisation dedicated to improving the public understanding of science. Tyndall carried out experiments at the Royal Institution which showed that carbon dioxide and water vapour trap heat in the lower atmosphere. I cannot match Tyndall's own beautifully lucid explanation that 'as a dam built across a river causes a local deepening of the stream, so our atmosphere, thrown as a barrier across the terrestrial rays, causes a local heightening of the temperature at the Earth's surface'. This local heightening is the greenhouse effect.

As a result of ongoing debates about global warming and mankind's role in it, we have become used to thinking of the greenhouse effect as a serious problem – but, without it, temperatures would be too low for liquid water to exist and our planet would be uninhabitable. The average temperature of an air-free planet is easy to calculate since it depends only on how much radiation is received from the Sun and what fraction of that radiation is reflected back into space rather than absorbed. If that calculation is performed for the Earth, the resulting prediction is that our average temperature should be about −18°C. Fortunately for us, the Earth's mean temperature is actually about +15°C and that 33°C difference is all down to the greenhouse effect. However, it is not inevitable that a planet will have any greenhouse effect at all, even if it has a thick atmosphere. In the same way that water is transparent to light and custard is not, only some gases are opaque to infra-red. The magnitude of the greenhouse effect therefore depends not only on the thickness of the atmosphere but also its composition. Oxygen and nitrogen produce almost no greenhouse effect and these two gases form most of the Earth's atmosphere. Carbon dioxide, water vapour and methane, on the other hand, are very effective greenhouse gases and there is enough of these around to make our atmosphere infra-red-opaque and our world hospitable to life.

Hopefully, you can now begin to see that acid rain falling onto the granites of Dartmoor, and elsewhere in the world, might have a direct effect on our climate since it is taking a greenhouse gas – carbon dioxide – out of the atmosphere. I'll have much more to say about Dartmoor later in this chapter but, at this point, I need to say a bit more about global warming due to carbon dioxide. Of the three greenhouse gases mentioned above, water vapour is the one that gives the biggest greenhouse effect and, in comparison, changes in carbon dioxide levels affect temperatures by a relatively small amount. For example, the increase in this gas produced by mankind in the last few hundred years has trapped enough extra heat at the Earth's surface to raise temperatures by only one third of a degree centigrade. So why am I making a fuss about carbon dioxide in this chapter and, rather more importantly, why are the majority of climate experts so concerned about mankind's production of it? There can be no dispute that CO_2 is currently building up in our atmosphere. Fifty years of continuous measurements of the air's composition, as well as measurements of air bubbles trapped in ancient ice sheets, show that the concentration is now 35 per cent higher than it was before the Industrial Revolution. There is also little room for doubt that this increase is due to human activities such as fossil fuel burning, deforestation and cement manufacture. The most straightforward evidence that this extra carbon dioxide has been generated by humans is simply that the amount of increase is roughly the same as that released by these activities. In fact some is missing, so, if anything, carbon dioxide levels should have gone up more than they have.

However, the increased greenhouse effect from this build-up of carbon dioxide has *directly* warmed the Earth by only 0.34°C. This calculation is based on well-understood physics backed up by detailed laboratory measurements and confirmed using direct measurements from spacecraft of the 'missing' infra-red radiation. The estimate that global temperatures should have risen by about one third of a degree centigrade is so well supported by such evidence that it is generally accepted both by sceptics of the human-induced global warming hypothesis and by those who support it. Given this widely accepted physics, even the expected doubling of carbon dioxide levels over the next century would raise temperatures by only 1°C and so it could be argued that avoiding such a rise would not be worth the staggering economic cost of taking the necessary action. However, that is not the end of the story.

There is a serious debate to be had about anthropogenic (i.e. man-made) global warming, but it is not about whether the carbon dioxide increase was produced by people (it almost certainly was) or about the immediate warming effect of this pollution (it is small). Instead, it concerns what happens when this minor initial warming triggers other changes such as increased cloud cover, increased water vapour and the melting of ice caps. These changes also affect global temperatures; some by reducing the overall warming (negative feedbacks) and some by adding to it (positive feedbacks). A good example is ice-albedo (i.e. ice reflectivity) feedback. If the Earth becomes a little cooler for some reason, then we'd expect the Earth's ice caps to grow. And, because they are very white, the enlarged ice caps will reflect some of the Sun's heat back into space and so the Earth will get cooler still. Conversely, if the world warms and the ice caps shrink, then the Earth will absorb more solar radiation than before and the warming will be enhanced. Hence, the ice-albedo effect produces positive feedback, a tendency to magnify any heating or cooling produced by other factors. There are many such feedbacks in the Earth's complex climate system and some magnify while others moderate any climate changes that may be occurring. As a consequence, the warming effect of increased carbon dioxide levels will be altered from the value expected based purely on the amount of extra heat trapped by the carbon dioxide acting alone.

The sizes of these feedbacks are notoriously difficult to calculate and so there is much room for debate over whether we need to worry about anthropogenic global warming. This debate has become one in which there is a great deal of mistrust, misrepresentation and counter-productive anger. Perhaps that isn't surprising since the stakes are so very high. If climate change sceptics are right, then society is being asked to make almost unsupportable financial sacrifices at a time of enormous economic difficulty for no good reason, and car enthusiasts will be forced by nothing more than political correctness to swap the throaty roar of their Ferraris for the rattle of the bicycle chain. On the other hand, if those who support the hypothesis of significant anthropogenic global warming are right, then our failure to make those same financial sacrifices will lead to permanent economic and environmental impoverishment of the whole planet.

As a consequence of the resulting atmosphere of suspicion, I feel obliged to say a few words about my own background and attitude before proceeding. I am an academic scientist, which means that I have the great good fortune to be paid to teach and to undertake research in a university. The majority of my published research has been concerned with computer modelling of how rocks form and how rocks deform. I'm therefore in a good position to understand what climate modellers do, since the computation techniques I use are very similar to theirs. But I am not a climate modeller myself and I have never received any funding for climate research. Indeed, nearly all my research funding over the last two decades has come from the oil industry – which, if anything, gives me an incentive to downplay the importance of climate change. Furthermore, I'd also have to admit that, like every scientist I know, if I could find a major flaw in the climate scientists' arguments I'd submit a paper on it within days to the most prestigious journals in the world because my scientific reputation would be made! Sadly for my chances of winning a Nobel Prize, I haven't found any serious flaws yet. But let me tell you what I have found.

To predict the full effect of changing greenhouse gas concentrations we need to know the 'climate sensitivity' of the Earth. Climate sensitivity is defined as the amount by which global *temperature* increases as the result of a small change to the *heating* of the planet. If there are no feedbacks then this is easy to calculate: it is about two thirds of a degree centigrade for each 1 per cent increase in heating at the Earth's surface. Doubling carbon dioxide levels would increase ground-level heating by about 1.5 per cent and therefore directly increase average temperatures by about 1°C. To take things further, though, computer simulations of the interactions between the Earth's atmosphere, oceans, cryosphere (ice cover) and biosphere are needed. These computer models can calculate the climate sensitivity including feedbacks. However, because of their almost incomprehensible complexity, the programs are vulnerable to the criticism that they may be flawed representations of the real Earth. I don't believe the models are badly flawed, but such concerns should not be lightly dismissed when they are raised by those who ask reasonable questions about whether anthropogenic climate change is as serious as most experts think. As many have said, these are multi-trillion-dollar questions and they require us to be more certain of our ground than is usual for purely scientific matters. We therefore need another way to move the debate forward so that we don't descend into pantomime arguments about whether computer modelling is helpful ('Oh yes it is!') or not ('Oh no it isn't!').

Fortunately, this model-complexity problem can be circumvented by using observations of the real climate system to independently verify the model results. For example, the increase in global mean temperature, from the 35 per cent increase in carbon dioxide levels over the past 200 years, is actually double the size predicted when feedback is ignored. This mismatch is data that can be used to estimate the strength of feedback, and hence climate sensitivity. The result is a climate sensitivity of 2°C for a doubling of carbon dioxide levels. It has been suggested that the measured temperature rise results from factors other than greenhouse gas changes. In particular, the suggestion that the rise results from changes in solar activity has been widely discussed and investigated in great detail. But, of all the different ways that interested scientists have attempted to explain away the relatively large temperature rise, the hypothesis that it results from strong positive feedback is the only one that does not require major revisions to our understanding of the climate system. Hence, based on purely observational evidence, the scientifically most conservative conclusion is that the Earth's climate system, as a whole, produces a net positive feedback: it significantly amplifies any direct warming caused by a build-up of greenhouse gases.

The occurrence of ice ages in Earth's recent past also supports this conclusion of significant positive feedback, because without high climate sensitivity, it is hard to explain why ice ages happen at all. In fact, ice age studies and more detailed studies of recent climate change both suggest that my simplistic calculation of climate sensitivity is too small. The likely reason for the discrepancy is that, in addition to raising carbon dioxide levels, human pollution has slightly increased the Earth's reflectivity, thus bouncing some of the Sun's heat back out into space. If it hadn't been for this unintended but benign consequence of pollution, the temperature rise would have been greater than the 0.7°C observed and my calculated climate sensitivity would have been larger. My calculated sensitivity is also too small because it takes no account of the thermal inertia of the oceans; the world's seas soak up a lot of excess heat and this delays the warming. When these factors are taken into account, then widely accepted values of climate sensitivity suggest that global average temperatures may climb as much as 3–6°C as a result of the doubling of carbon dioxide levels expected during this century – and those are temperature changes not seen since the end of the last ice age. For me, the message is clear: there are great uncertainties in the prediction of future temperature changes but even very optimistic assumptions predict major difficulties ahead. Wishful thinking is not a good strategy for dealing with this.

The politically contentious issue of anthropogenic climate change may seem a long way from my book's central theme of what makes planets habitable, but there is a strong connection between these apparently very different subjects. The level of climate sensitivity revealed by research into present-day global warming is hard to reconcile with the surprisingly pleasant climate of our planet in the distant past when the Sun was fainter. Growth in solar heat output has been going on throughout the Earth's history and, if the climate sensitivity discussed above was accurate throughout that time, then 500 million years ago temperatures should have been more than 10°C cooler than they are today. But, as I will discuss in the next few chapters, the evidence in the rocks points to a world that, half a billion years ago, was at least as warm as it is today and probably warmer. Furthermore, climate sensitivity calculations for 4 billion years ago, when life first emerged and the Sun was 30 per cent less luminous than today, predicts temperatures far too cold to allow liquid water or life anywhere on our planet. Nevertheless, life did emerge and ancient rocks show clear evidence of flowing water. Early Earth managed to be a much warmer place than we should reasonably expect and, as a result, life was able to arise and thrive far sooner than it would otherwise have done. The problem of explaining the contradiction between the solar physics and the geological evidence is known as the 'faint young Sun paradox' and, if our planet hadn't somehow solved this paradox, we wouldn't be here to wonder why.

Part of the explanation for this surprisingly warm ancient world can be found by returning to the Burrator quarry on Dartmoor, and the effect of acid rain falling on granite. As mentioned earlier, the rain is acidic because it has absorbed a little carbon dioxide from the air and so rainfall should slowly scrub the atmosphere completely clean of the gas. The main reason it doesn't do so is because the gas normally bubbles slowly back out of streams, rivers and oceans to return to the air from which it came. However, when the acid rain happens to dissolve volcanic rock such as Dartmoor granite, that bubbling out no longer occurs because the chemical reaction transforms the carbon dioxide into a more permanent form, bicarbonate, and also loads the water with metallic ions such as iron, magnesium or calcium. Hence, the chemical reaction between rain and rock has locked the carbon into water, which finds its way to the sea. The overall effect is that carbon originally located in the atmosphere is now dissolved in the oceans. But the story doesn't end here. To show what happens next, I'd like to describe a very different quarry in another part of the UK.

My memory is a little vague on the details as I haven't been in the quarry for nearly 40 years. The only reason I remember the place at all is because this is where I collected my own first fossils. The quarry was a ten-minute cycle ride from my home in the small town of Beith, in western Scotland. A dry-stone wall had been built around it to keep small boys like me out, but it was ineffective against the lure of the quarry and its shallow mine-workings. This attraction was massively enhanced by the fact that the loose rocks in the quarry were packed with fossilised fish bones. Or so I thought. My colleagues today will be horrified to hear me admit this, because those fossils were actually the remains of crinoids, which are about as different from fish as it is possible for another animal to be. Crinoids are invertebrate animals, related to starfish, which live by filtering food from seawater. To feed effectively, they are anchored to the shallow sea floor by stalks that keep them stationary while water currents sweep edible debris past. These stalks give crinoids the superficial appearance of plants and, as a result, they are sometimes called sea lilies. Like all animals, crinoids eventually die through mischance, disease or old age and, when they do, the soft tissues holding the stalks together rot away to leave behind the disarticulated remains of their hard parts. These bear a very superficial resemblance to bones, hence my schoolboy error.

Crinoids are relatively rare today but when my fossil crinoids were alive, during the Carboniferous Period 320 million years ago, they were common and carpeted much of the shallow, warm seas that then flooded most of Britain. These seas extended from the active volcanoes of south-west England (where the granites of Dartmoor were to form 30 million years later) to the high mountains of southern Scotland. North of these mountains lay a rift, the Midland Valley of Scotland, where the land was rapidly subsiding and it was here that my fossils were formed. In Carboniferous times, Britain was not the cool, wet place it has become today. As a consequence of plate-tectonic-driven continental drift, Britain then lay near the equator and the Midland Valley was usually a hot, arid desert. However, episodic climate change brought occasional much wetter conditions when the valley transformed into a tropical swamp inhabited by giant dragonflies, early reptiles and tree-sized ferns. The peat left behind in these swamps later became the coal-beds of central Scotland. On yet other occasions, when land subsidence or sea-level rise was fast enough, the valley became inundated by the sea. Each of these flooding events lasted many thousands of years during which hundreds of generations of crinoids and other organisms lived, died and left their remains behind to form layers of rock tens to hundreds of metres thick. It was during one of these episodes that the rocks containing my crinoids were laid down.

The crinoid-packed stones I picked up as a schoolboy, along with similar rocks from across the world and of all ages, are therefore made from the harder bits of long-deceased creatures. The type of dead animals to be found will vary with time and place, so that some deposits contain crinoid fragments while others consist of ancient corals, the remains of prehistoric plankton or even fossilised bacterial mats. Wherever they are, and whenever they formed, all such rocks are composed from the mineral calcium carbonate and are therefore called carbonate rocks or, more often, limestone. This calcium carbonate was originally produced when the marine creatures extracted calcium ions and bicarbonate ions from seawater to make their hard parts and these are the very same chemicals that were dumped into the sea by the acid-rain weathering of ancient places similar to modern Dartmoor. The overall result of acid-rain weathering followed by limestone deposition is therefore that carbon, originally contained in atmospheric carbon dioxide, has first been incorporated into bicarbonate ions dissolved in the sea and then transformed into calcium carbonate locked up as sediment on the sea floor. Climatic, geological and biological processes have combined to suck carbon dioxide from the air and to spit it out as solid rock.

Weathering of volcanic rocks by acid rain is a worldwide process that is particularly vigorous in the warm and wet tropics. This weathering has been going on since our planet was young and, even before carbonate-producing organisms became wide-spread, limestone precipitated on the sea floor once weathering-generated ions became concentrated enough in the seawater. Carbon dioxide has therefore been continuously taken from our atmosphere, and added to carbonate rocks, over almost the entire lifetime of our planet. The gas has now become very rare, with a concentration in the air of less than 0.04 per cent. This level is so low that it is starting to make life difficult for plants, all of which need carbon dioxide to exist. As a result, an entirely new class of super-efficient vegetation that can cope with such low levels has evolved in the last 30 million years. Grasses are the most familiar of these, and they are extremely good at gleaning carbon from the depleted reserves remaining in our atmosphere.

However, limestone creation has not yet completely removed carbon dioxide from our atmosphere because it is partly replenished by the reverse chemical reaction: the break-down of carbonate to produce carbon dioxide. There are two ways in which this happens. The first of these involves rainfall yet again because, if you'll forgive me for corrupting a biblical quotation, 'the rain falls on volcanic rocks and carbonate rocks alike'. As we've seen, acid rain on volcanic rocks traps carbon dioxide but acid on limestone has the opposite effect – it liberates carbon dioxide. Indeed, one way we teach students to recognise carbonate rocks makes use of this reaction. Dribbling dilute hydrochloric acid on limestone makes it fizz like a newly opened bottle of cola as carbon dioxide escapes from the rock to bubble back into the atmosphere it left millions of years before. The effect is quite dramatic and many students will dab a splosh of acid onto any new rock they see on the off-chance it might be limestone, just for the fun of watching it fizz. A rather slower release of carbon dioxide into the air happens when acid rain, rather than a student, attacks limestone but as this goes on day after day and year after year, the overall effect is significant. You can gain a good idea of the cumulative effect of acid rain by looking at the text on old limestone monuments; they become unreadable over the centuries as the acid slowly eats away at their surfaces. Acid rain falling on larger carbonate objects, such as entire mountains of limestone, generates millions of tonnes of carbon dioxide every year and this helps to stop the atmospheric level from falling to zero.

The second way in which limestone rocks are broken down to produce carbon dioxide is slightly more complex. The huge tectonic forces that, in a few hundred million years, have moved the UK from the equator to its current chillier location can also subduct surface rocks into the deep Earth as I discussed in the last chapter. For example, subduction is currently happening off the western coast of South America where the Nazca plate (a part of the Pacific) is moving towards South America at a speed of about 7 centimetres per year. About 2 centimetres of this is absorbed as the continent crumples to push up the Andes mountain chain but the remaining convergence is taken up by the Nazca plate slipping underneath South America and into the deep Earth. Subducted rocks heat up in the hot interior of the Earth and the resulting magma percolates back up to the surface to produce volcanoes. If subduction and heating happens to limestone rocks, they give up their carbon dioxide, and this is carried to the surface in the magma and emitted by the volcanoes as they erupt.

Carbonate rocks can therefore be destroyed and their carbon put back into the atmosphere, but fortunately, much of the limestone produced over the lifetime of our planet has avoided being eaten by rain or pulled into the fiery depths of the Earth. The quantity of carbon stored in limestone rocks – all of which was in the Earth's atmosphere when our planet was young – is therefore truly staggering. This high level of greenhouse gas in the young Earth's atmosphere was lucky for us because the early Earth under a faint young Sun would otherwise have been uninhabitable. Since that time, the amount of carbon in our atmosphere has gradually dropped so that a falling greenhouse effect has continuously compensated for the increasing heat supply from the Sun as our star and planet have grown old together. I hope that you can see, from all this, that the idea that the Earth has had a constant climate for billions of years simply because nothing much has changed, is extremely naive. In reality, solar heating and atmospheric composition have continuously changed throughout our planet's history and, in principle, these changes could have led to catastrophe. But they didn't.

We can see how bad things could have been by looking at our sister planet Venus, which still has a thick, carbon dioxide-rich atmosphere. Venus is closer to the Sun than the Earth is, and is unsurprisingly a warmer world than ours. However, the Russian Venera probes sent to Venus in the 1970s and 80s showed that it is more than just a little warmer. The first of the series, Venera 7, landed on Venus in December 1970 and its electronics survived the searing heat for 23 minutes – long enough to send back the first direct measurement of the surface temperature. At 465°C, the surface of Venus is hot enough to melt lead and boil sulphur.

These extremely high temperatures were a bit of a surprise, because if the Earth were to be moved to the same distance from the Sun as Venus, the increased heat would warm our planet by only about 50°C. Of course, there would be some knock-on effects from this warming. We might, for example, expect the Earth to get cloudier as more water evaporated from the warmed seas and this would surely affect the climate, too. Venus is indeed very cloudy, which is why it looks so bright from the Earth, but these clouds reflect light and heat away. One of the effects of clouds is therefore to produce cooling, and putting this into the calculation predicts a temperature of about 10°C for our transported Earth. Thus, if proximity to the Sun and planetary reflectivity were the only factors that mattered, Venus would actually be colder than the Earth. However, there is one final factor that has to be taken into account: the greenhouse effect from Venus's thick, carbon dioxide-rich atmosphere. The amount of carbon dioxide in the Venusian atmosphere is similar to the amount there would be in Earth's atmosphere if all the carbon in all the limestone rocks of the world were suddenly put back into the air. The resulting massive greenhouse effect gives Venus its high temperatures. But why should there be so much more free carbon dioxide on Venus than on Earth?

The difference between the Earth and Venus is that our neighbour has lost its water and retained its carbon dioxide while, as we have seen, the Earth has done the exact opposite. The environment on Venus was not always so hostile but, as computer models show, when Venus was younger and just a little warmer than the Earth, ferocious storms transported water vapour high into the atmosphere where solar radiation and the solar wind broke the vapour into its constituent hydrogen and oxygen. The hydrogen then escaped into space, thereby removing for ever a large fraction of Venus's water. The little water that was left reacted with sulphurous emissions from volcanoes to form sulphur dioxide-rich clouds. Today, Venus is not a nice place. It has clouds with the composition of concentrated battery acid, surface temperatures that would melt lead, and no rainfall. The lack of rain is particularly critical, as this means there are no seas and so no easy way to produce limestone to get rid of all that troublesome carbon dioxide.

Coming back to Earth, it seems an amazing coincidence that the slow loss of carbon dioxide from the atmosphere has closely compensated for the gradual increase in solar heating. This picture is a little over-simplified because methane, another important greenhouse gas in the atmosphere, has also declined throughout Earth's history. Nevertheless, there may be a simple reason why carbon dioxide levels have fallen at the right sort of rate. On a warmer, wetter Earth there is more volcanic rock weathering. Hence, if the Earth happens to warm a little, carbon dioxide is removed from the atmosphere faster than it is replenished. Carbon dioxide levels then fall, greenhouse warming decreases and the initial warming effect is greatly damped. If the Earth cools, the opposite happens: weathering decreases, carbon dioxide builds up and greenhouse warming increases. The existence of this negative feedback is the reason I have claimed that the Earth's climate system is thermostatically controlled. The process is slow and takes millions of years to correct any cumulative drift of our world's temperature upwards or down, but it is an important component of our climate that has helped prevent potential disasters such as those that have befallen Venus and Mars.

Thus, on the Earth, a complex set of chemical reactions has transformed volcanic rocks in the presence of atmospheric carbon dioxide to produce limestone and this has reduced the long-term climate sensitivity of the Earth. Without this the Earth's temperature history would certainly have been far less stable than it was. However, most researchers believe that the effect has been to reduce, rather than eliminate, an overall positive feedback in our climate system. Biological, geological and astronomical processes that have warmed, or cooled, our planet have therefore still had their effects magnified by the Earth's climate system. Furthermore, the long-term temperature history of our world is, surprisingly, one of gradual fall in temperature despite the steady rise in solar output. This cannot be explained as resulting from negative feedback since this would, at best, reduce the expected warming rather than put long-term climate change into reverse.

None of this looks like the behaviour of a well-designed climate control system. If this was a heating and air conditioning system in a house you would have to cross your fingers and hope for mild winters and cool summers; the house has its wires crossed so that the air conditioning comes on in the winter and the radiators are warm in the summer! Fortunately, we have had the equivalent of mild weather for the vast majority of our planet's history, but, as we'll see in the next chapter, there were odd occasions when conditions were severe enough to seriously damage the biosphere. Bad things have happened, just not very often.

6

Snowballs and Greenhouses

Imagine a time when carbon dioxide levels in the atmosphere have doubled. Methane, released from warmed sea floor in the oceans as well as from melted permafrost on land, has also accumulated in the atmosphere and enhanced the global warming so that the Earth's average temperature has risen by 8°C. The resulting disaster is particularly acute inside the Arctic and Antarctic circles where temperatures have risen by 20°C and the ice caps have completely melted away. Our planet had been in the grip of an ice age in the geologically recent past but now has no sea ice at all. Catastrophe has come to the tropics as well as to the poles. There had been deserts to the north and south of the equator before climate change began but these harsh, sparsely inhabited zones were separated by a wide, lush band of tropical vegetation straddling the equator. That verdant, rich band of life has been squeezed out of existence by greatly expanded deserts that have coalesced. Half the entire surface area of the continents is now arid and bare. Matters are, if anything, even worse in the seas than on land. Currents, normally driven by the sinking of cold, dense polar seawater, have stagnated so that life in the deep oceans is dying from lack of oxygen. Animals and plants experience an unprecedented mass extinction as species encounter extremes of heat and drought on land and oxygen-starvation in the oceans. Ninety-five per cent of marine life and 70 per cent of terrestrial species become extinct. Armageddon comes because too much carbon dioxide has been dumped into the air.

This horrific vision is not from 250 years into the future.

This is an ancient global warming catastrophe from 250 million years in our past. The carbon dioxide was released by huge volcanic eruptions that covered much of present-day Siberia in a layer of solidified lava and volcanic ash several kilometres thick, and the resulting temperature rise was enough to wipe out nearly all multi-celled species. It would be hard to find clearer evidence that complex organisms are vulnerable to even quite moderate climate change. Perhaps the best known animals to disappear at this time were trilobites, distant relatives of modern horseshoe crabs. Trilobites had successfully dominated the seas for nearly 300 million years but nothing had prepared them for rapid global warming and they were completely exterminated. The environmental changes were simply too fast to allow adaptation by these relatively large, slow-breeding organisms. The species that vanished were eventually replaced by many entirely new ones but this took a long time and Earth's biodiversity was measurably reduced for 10 million years after the catastrophe. The dramatic change in dominant life-forms revealed in the fossils from this time has led geologists to assign different names to the Periods either side of this event. The preceding 50 million years is the Permian Period and the succeeding 40 million years is the Triassic Period; the disaster is therefore known as the Permo-Triassic mass extinction event.

The pattern of death and, eventually, new life seen in the Permo-Triassic mass extinction was repeated in each of the other, similar catastrophes that have hit our planet five times in the past half-billion years. In each disaster a substantial fraction of existing species died out to be replaced over the next 5 to 10 million years by new animals and plants that evolved from the survivors. Extinctions on each occasion were caused largely by bad weather, although the exact causes of the climate change varied from one catastrophe to the next. The best known example is the mass extinction 65 million years ago at the end of the Cretaceous Period. In that case there seems to have been a volcanically induced gradual deterioration in the climate followed by a coup de grâce of a major meteorite impact that would have had additional severe climatic effects. The result was the extinction of the dinosaurs, along with many other life-forms, but the subsequent opportunities opened up for new species allowed mammals to become the dominant large animals of the present day. Occasional disasters therefore encourage evolutionary innovation, but you can have too much of a good thing. If major climate catastrophes happened every few million years there would be insufficient recovery time between the successive mass extinctions and biodiversity would be permanently reduced. Fortunately, on Earth, the average gap between the really big disasters has been about 100 million years and life has been able to recover and thrive after each one.

There have therefore been quite a few mass extinction events but the Permo-Triassic was the biggest one to hit life in the last 542 million years. This may seem a slightly odd statement. Why not count catastrophes over, say, the last 1 billion years or, more sensibly still, over the entire 4.5 billion-year lifetime of our planet? In fact, 542 million years is not an arbitrary choice. This Eon (geo-jargon for a very long period of time) is called the Phanerozoic, 'the time of visible life', because it's only during this most recent half-billion-year stretch of time that animals have existed with easily fossilised parts such as skeletons, exoskeletons and shells. So, the Permo-Triassic event is simply the biggest mass extinction we have clear evidence for. We have too few fossils from further back in time to be able to tell whether earlier life suffered even worse setbacks. However, it is likely that more severe disasters actually did occur in the distant past because climate variations seem to have been more extreme in those ancient times. The Eon preceding the Phanerozoic is called the Proterozoic, 'the time of earlier life', and this Period, which began 2.5 billion years ago, experienced the most savage temperature swings the world has ever seen.

Evidence for Proterozoic climate change dwarfing that at the end of the Permian is found at many places around the world such as Canada, Brazil, Namibia, Svalbard, Australia and Oman. I've recently been to see these clues for myself in yet another location: Shetland.

The Isles of Shetland are a bleakly beautiful group of islands so far north of Britain that they are closer to the Arctic than they are to London. Shetland's many peninsulas and interspersed sea lochs give these islands a characteristically intricate coastline that was formed, much like the Norwegian fjords, by the action of glaciers during the Earth's current ice age. But the story of a much earlier and more extreme ice age is also written into Shetland's rocks. On the peninsula of Strom Ness I have placed my hand on sediments that show spectacularly rapid and dramatic climate change. In a photograph I couldn't resist taking, the index finger of my left hand rests on sediments deposited during a time of near-global glaciation while my ring finger sits over carbonate rocks laid down in a world far warmer than our own. The change-over, hidden by my middle finger in the picture, was abrupt but with no sign of any break in deposition. It looks as if the climate switched from bitterly cold to stiflingly hot almost overnight or, at least, over thousands rather than millions of years.

So, what's the evidence in these rocks for the dramatic climate change I've claimed? I'll come back to the carbonates and how they relate to a very warm world later, but, for now, let me concentrate on the cold-climate sediments preserved at Strom Ness. These are covered in variegated patches of lichens, which, while helping them to stand out from the relatively lichen-free carbonates, makes it hard to see the distinguishing features of deposition in chilly times. However, after rooting about for half an hour or so, I found what I was looking for. On a relatively lichen-free area I could see that the rocks were made from very fine grains arranged in beds typically a millimetre thick or less.

This can happen only if the sediments forming the rock were laid down in a quiet environment where even small grains could settle out to rest on the sea floor and where the thin beds thus formed would not be stirred up and disarranged by waves, currents or tides. But, in contrast to this evidence of tranquillity, I could also see a pebble that appeared to have been dropped onto this ancient sea floor from a great height. The beds below the pebble were clearly deformed by the impact and, in any case, the surrounding sediments showed that there were no currents capable of moving such a relatively large object. The pebble seemed to have fallen out of the sky. Fortunately, there is a simple explanation for such an apparently unlikely thing. The pebble probably dropped off the bottom of an iceberg, since there is no other known, natural way of transporting isolated large stones into calm water. Of course, one dropstone on Shetland does not a snowball Earth make, but rocks of similar age in other parts of the world show even more dramatic examples with dropstones sometimes a metre or more across sitting in otherwise fine sediment. Many other types of glacial sediment have also been seen all over the world from rocks of this age, and some of these cold-climate rocks are found in places known to have been near the equator at the time. The Earth must have been a very cold place 635 million years ago and may even have been completely frozen over.

Computer climate models show that an almost completely frozen Earth is not only possible but actually a likely outcome once polar ice caps grow to within about 30 degrees latitude of the equator. A runaway cooling then sets in where the increased ice cover reflects heat into space, causing further cooling and yet more ice until the whole world is white, beautiful and all but dead. In these models the Earth becomes a permanently frozen globe with an average temperature of −50°C. This prediction remains controversial, however, with many geologists arguing for a slushball rather than snowball Earth, but that is just an argument about whether equatorial temperatures were lethally low or merely bitterly cold. From my point of view the important and undeniable message from the rocks is that the Earth had a very different, and much colder, climate 635 million years ago than it does today.

So, what of the carbonate rocks at Strom Ness, which tell a dramatically different tale of an extremely warm Earth? Similar 'cap carbonates' are found on top of late-Proterozoic glacial deposits at many locations around the world and they record the spectacularly rapid end of global glaciation. A cold world is a very dry world with most water locked up in ice and too little heat to allow much evaporation into the atmosphere. On snowball Earth it never rained and rarely snowed; like C.S. Lewis's mythical land of Narnia it was always winter but never Christmas. As a consequence of this lack of rainfall, the removal of carbon dioxide from the atmosphere by acid rain, as discussed in the last chapter, ceased; but since volcanoes continued to emit the gas, it built up in the atmosphere. After about 4 million years there was enough greenhouse warming from this to allow equatorial ice to begin melting. The process that drove the Earth into a snowball state then went into violent and rapid reverse. The disappearing ice reduced the reflection of heat into space, giving further warming and even less ice. This became a runaway process with melting driven at an ever-accelerating rate until all the ice had gone.

With the rising temperatures, highly acidic rainfall falling out of the carbon dioxide-rich atmosphere onto volcanic rocks produced the calcium carbonate in the sea that was needed to generate the thick, globally extensive limestone seen in Shetland and elsewhere. Within a few thousand years of the thaw's beginning, ice completely disappeared from the Earth and our planet entered a super-greenhouse phase in which global mean temperatures may have risen to as much as 50°C above freezing and only slowly returned to more normal levels as the carbon dioxide concentration gradually fell. Thus, in a geological instant, temperatures possibly rose by as much as 100°C. This makes the 8°C change in the Permo-Triassic mass extinction event look like a very minor incident – and this snowball-to-greenhouse upheaval happened not just once but at least four times during the late Proterozoic. Furthermore, a similar phase of repeated snowball Earth episodes also happened in the early Proterozoic around 2 billion years ago. The fine details of all this Proterozoic climate instability remain highly controversial and many experts would disagree with much of what I have said in the preceding few paragraphs but, nevertheless, it remains incontrovertible that our planet was capable in the past of exhibiting very severe climate change; climate change bad enough to wipe out most life-forms and leave behind an unimaginably impoverished biosphere.

In the last half-billion years, temperatures have become more consistent and, for these more recent times, there is less room for debate about how much climate change really occurred because the temperature history is written in the fossil record. That may seem strange. Bones and shells fossilise but softer parts usually do not. The claim that temperature fossilises therefore seems quite outrageous, but, thanks to my clever geochemical colleagues, it really is possible to estimate ancient temperatures. Advances in recent decades, in both the precision with which the chemical composition of ancient rocks can be analysed and in understanding what these analyses can tell us about ancient environments, have been extraordinary. This is the Earth-science breakthrough I alluded to in Chapter 1 and it has provided us with deep new insights into the history of our planet. The chemical fingerprints of ancient seawater temperatures left in the shells of ancient sea creatures are an excellent example of this, and they result because the water incorporated into the carbonate of these shells became heavier when the weather was colder. Let me expand on that.

You may well have heard of heavy water and know that the Nazis attempted to manufacture large quantities of it during the Second World War using cheap Norwegian hydro-electricity. This was a major part of their fortunately unsuccessful nuclear weapons programme. In heavy water, all the hydrogen in the water molecules has been replaced by deuterium, a rare form of hydrogen that is twice the normal weight. This deuterium is an example of what is called an isotope, and elements other than hydrogen have isotopes too. In particular, the heavy water that interests climate scientists is quite different to that used by nuclear engineers because it is made with heavy oxygen rather than heavy hydrogen. Oxygen usually comes in a form called oxygen-16 (which means it has eight protons and eight neutrons in its nucleus), but one atom in every 500 is oxygen-18 (with two extra neutrons). Heavy water containing oxygen-18 occurs quite naturally. In fact, your body holds about 100 grams of it. Glaciers, on the other hand, contain hardly any of this oxygen-heavy water at all. The extra weight of these water molecules that contain oxygen-18 makes them harder to evaporate and also more likely to return quickly to the surface as rain. So, by the time water has been evaporated from the tropical seas and transported in clouds to the poles, it contains little of the heavy oxygen. Ice caps are therefore made almost entirely from what you could call light water. The overall process is quite similar to that used for making strong alcoholic drinks from liquids containing less alcohol. In the same way that evaporation and re-condensation of fermented liquid concentrates the easily evaporated alcohol, so evaporation and condensation of sea-water concentrates the easily evaporated light water. The Earth acts like a large whisky distillery, although sadly the products are not quite as tasty.

This distillation of seawater can be used to unravel ancient climate because the heavy, oxygen-18 containing, water molecules become a little more concentrated in the sea when there is lots of the light water locked up in ice caps. If this heavier seawater is then incorporated into growing shells, the variations in oxygen-18 content over time become preserved. In fact, the variation is further enhanced since organisms take up oxygen-18 more efficiently when the water is cooler. There are also complications resulting from the more acidic nature of the ancient oceans when carbon dioxide levels were higher, because this acidity also affects the efficiency with which organisms absorb the heavier isotope of oxygen. However, once all these effects have been disentangled, what emerges is an estimate of equatorial temperatures throughout the 542 million-year period that shelly animals have existed on Earth.

Equatorial temperatures vary less than the Earth's overall mean temperature because polar temperatures always go up and down more strongly than equatorial ones. The results therefore need to be amplified a little to give the global picture, but what emerges is the very robust conclusion that, during the Phanerozoic, temperatures oscillated up and down three or four times with an amplitude of about 10°C. In addition there may have been an overall cooling trend of a few degrees, although this is still being debated. This picture of oscillation around a roughly constant temperature is quite different to what we'd expect based on the climate sensitivity models discussed in the last chapter. Those models predict that temperatures should have climbed during the Phanerozoic by around 10°C as a result of the steadily warming Sun. During the last half-billion years the cooling consequences of geological and biological evolution must therefore have compensated for the warming due to solar evolution. Evidence from further back in time is more contentious but, nevertheless, what there is reinforces the message. When it comes to climate change, Earth evolution-driven cooling has matched, or even outpaced, solar evolution-driven warming.

The other long-term trend that emerges from these analyses

is that fluctuations in climate seem to have become less severe in the last half-billion years; we no longer have snowball Earth episodes. But can we be sure that the Earth has outgrown the adolescent behaviour of the Proterozoic and become a more mature, reliable and predictable planet in the Phanerozoic? The modern world is a very different place from the one recorded in the Shetland rocks, and there seems to have been a fundamental change on the Earth 542 million years ago when we entered the present Phanerozoic Eon. Life stepped up a gear at that time in what has been called the 'Cambrian explosion', the startlingly sudden appearance of widespread and recognisably animal life. As well as not having much in the way of animal life, the Proterozoic Eon differed from the Phanerozoic in that, even if we discount the snowball Earth events, it experienced much greater environmental change. At the beginning of the Proterozoic, oxygen levels in the atmosphere were a hundred times smaller than they are today and global mean temperatures may have been as high as 70°C. Noon-day temperatures at the equator must have approached boiling point! In contrast, by the end of the Proterozoic, oxygen levels and temperatures were broadly similar to those of today. For some reason the Earth's climate settled down half a billion years ago and many researchers believe that this is a direct result of the new, wonderfully rich and complex Phanerozoic biosphere controlling the climate through a network of negative feedback processes that did not exist in the simpler Proterozoic world. This idea has been called the Gaia hypothesis, and I'll look at it in detail later on.

However, it is possible that the Gaia hypothesis has simply confused cause and effect. Perhaps the complex Phanerozoic biosphere is a *consequence* of climate stability rather than its cause. If the Earth had had a more interesting climatic history – for example, occasional snowball episodes in the last few hundred million years – then our beautiful Phanerozoic biosphere would not have survived and we wouldn't be here to record the fact. Gaia is therefore not necessary to explain the observations, since the anthropic selection effect I discussed in Chapter 1 accounts just as well for what we see today. This doesn't mean that Gaia is wrong; just that there is an alternative explanation. To decide between Gaia and anthropic selection, we need to look a little closer at the causes of environmental and climatic change on Earth.

7

Staggering Through Time

The Earth's climate history has been like a drunken stagger home after a party; each step almost random but with an underlying directionality. Over the last 3 billion years, the global climate system has frequently gone in unpredictable directions but there has also been a distinct trend towards gradual cooling. This chapter takes a closer look at the causes of the lurching and likely reasons for the underlying cooling.

By far the most significant environmental change during this 3 billion-year period was the rise of oxygen. The massive increase in the atmospheric concentration of this initially highly toxic gas had profound consequences for all living organisms, but it also had serious climatic consequences. At the beginning of the Proterozoic our planet was at least as warm as today, and possibly much hotter, even though the fainter Sun produced only about 85 per cent of the heat it does now. This unexpectedly high temperature can only have been because the Earth was less reflective or because it had a bigger greenhouse effect. The most likely explanation is higher greenhouse warming due partly to higher carbon dioxide levels and partly to higher levels of atmospheric methane, which would have been produced in vast quantities by the kinds of bacteria that dominated the Earth at that time. However, high concentrations of methane cannot co-exist in the atmosphere with oxygen since they react together to produce water and carbon dioxide. So, it's quite likely that a falling greenhouse effect and rising oxygen levels are linked: as the oxygen concentration rose, methane reacted with it and levels of that gas fell. But where did the oxygen come from? That is a more complex story than you might at first imagine.

Oxygen is produced by plants during photosynthesis, the process by which they use the energy from sunlight to manufacture carbohydrates, the basic foodstuff of all life. However, plants evolved on Earth less than 500 million years ago and were therefore not the source of oxygen in the more distant past. The first photosynthetic organisms were bacteria, as we saw in Chapter 2, and they evolved even before the Proterozoic began 2.5 billion years ago. Carbohydrates, as the name implies, are made from carbon, hydrogen and oxygen. The photosynthetic microbes took carbon and oxygen from carbon dioxide dissolved in the sea but they also needed a source of hydrogen. Early photosynthetic bacteria got this from hydrogen sulphide gas emerging from volcanic vents on the sea floor. The sulphur atoms attached to this hydrogen were largely surplus to requirements and these organisms therefore excreted sulphur and sulphur compounds into the environment.

The shallow Archaean (the name given to this early Eon by geologists) seas must therefore have been painted in the vivid red, orange and yellow colours of sulphur and sulphur compounds, but 700 million years later, those seas suddenly turned green. Microbes had arisen that got their hydrogen fix from hydrogen oxide instead of hydrogen sulphide and they used chlorophyll, the green pigment found in all plants, to do so. Hydrogen oxide is better known as water, and although it is by far the most plentiful source of hydrogen on our planet, it takes a lot of energy to split this compound into its constituents. Water is a tough hydrogen source to use and it was a long time before any organisms appeared that could do it; but once the trick evolved, the bacteria using it thrived and started to dump the useless, to them, oxygen compounds into the environment. A little under 3 billion years ago the world became dominated by the first organisms to give off pure oxygen gas, cyanobacteria. These photosynthetic micro-organisms are very green and very successful; so successful that they are still common on our planet today.

The evolution of cyanobacteria led to the biggest and most deadly pollution incident in the whole of Earth history because free oxygen was lethal to all existing life. Indeed, this may be why micro-organisms initially evolved this chemical curiosity; it allowed them to kill off competitors. Evolution is full of examples of important innovations whose original purpose was quite different to their final one. Free oxygen emission may have originally evolved as a weapon in the constant chemical warfare between micro-organisms and only later became the basis of nearly all food manufacture on the planet. As we have seen, the free oxygen also had the potential to cause massive climate change. The climatic effects weren't too dramatic to start with. The very reactive oxygen combined with volcanically produced minerals, such as iron, dissolved in the sea and this kept atmospheric levels of oxygen down. Nevertheless, local oxygen levels in the shallow seas rose significantly and the life-styles of many micro-organisms become untenable. Fortunately, microbes are tough, rapidly evolving organisms and forms appeared that were oxygen-tolerant. Some bacteria simply learned to neutralise the oxygen but one group of purple bacteria evolved that used oxygen and carbohydrates as a source of energy. In other words, they ate and breathed as we do. Indeed, it is more accurate to say that purple bacteria still do all the eating and breathing since mitochondria, the oxygen-consuming power stations of plant and animal cells, are almost certainly descended from purple bacteria captured and enslaved for this purpose by the single-celled common ancestor of all plants and animals.

The evolution of breathing organisms provided yet another mechanism that kept oxygen levels low. All the oxygen released by photosynthetic organisms when they made carbohydrates was turned back into carbon dioxide when the photosynthesisers were eaten. Non-photosynthetic organisms, such as humans, get the energy they need for life by doing the reverse of photosynthesis. They combine carbohydrates with oxygen from the air and then breathe out carbon dioxide. Photosynthesisers therefore turn carbon dioxide into oxygen and then oxygen-breathers turn it back into carbon dioxide. Once breathing organisms had evolved there could be no increase in atmospheric oxygen levels; that is, unless some carbohydrate avoided being consumed so that the oxygen released during its manufacture stayed in the atmosphere. Given this, it's not surprising that the atmospheric oxygen level remained a tiny fraction of its modern value throughout most of Earth's history. The real puzzle is, why did it ever rise? That brings me to the fascinating topic of mud.

Mud doesn't sound fascinating to most people. I've heard the science of sedimentology described as 'mud moves from here to there and then from there to here', which is reminiscent of the classic description of history as 'just one damned thing after another'. Not very flattering, especially to those of us who have spent careers largely devoted to understanding how mud (and other detritus) is moved around by rivers, currents, waves and tides. My eldest son recently used similar language to dismiss this as the most boring of all possible topics: 'Mud gets washed into the sea, what's interesting about that?' Actually, with that simple statement he hit the nail on the head. Washing of mud into the deep ocean really is interesting! Without the settling of mud onto the deep sea floor, oxygen levels couldn't have risen in the atmosphere and large oxygen-consuming animals, like us, wouldn't exist. Mud contains organic material produced by photosynthesisers and some of it gets buried quickly enough in deep parts of the ocean to avoid being eaten. For each atom of carbon, extracted from carbon dioxide during photosynthesis and then buried at sea, there is a molecule of oxygen left behind in the atmosphere. The amount of carbon buried every year is a tiny fraction of the total biomass contained in plants, animals and other organisms, but over geological time it adds up, allowing oxygen levels to grow. Through the Proterozoic the burial rates started to rise, and as reactive gases and minerals were also used up, atmospheric oxygen levels increased. The resulting methane destruction then forced temperatures down. This trend seems to have continued into the Phanerozoic with events such as the rise of land plants acting to accelerate burial of organic-rich muds even further and push temperatures lower still. The world has never been colder, on average, than it has been during the last 30 million years (the snowball Earth periods were quite warm on average since the glaciations were relatively brief).

Declining greenhouse warming, countered by a steadily warming Sun, therefore explains the broad picture of long-term moderate climate change on our planet. But why did temperatures also fluctuate both in the Proterozoic and more recently through the Phanerozoic? Atmospheric composition and solar activity are not the only things that have changed significantly over the long history of our planet. Variations have also occurred in the Earth's reflectivity because of changes in ice, clouds, seas, continents and plants. Cloud cover, for example, must have changed dramatically through the ages. Cloud formation is strongly encouraged by atmospheric dust, and this dust has probably increased in volume because the Earth's continents have grown through time, and that's where dust comes from. More recently, the amount of dust must have changed significantly when the continents were conquered by plants 400 million years ago. In addition, plants encourage cloud formation by increasing the efficiency of evaporation, and cloud cover is also affected by biological activity in the oceans. Some ocean-going microbes produce droplets of chemicals in sea-spray that migrate into the atmosphere and help clouds form. So, the amount of cloud cover has probably gone up and down dramatically over the eons in response to continental growth and a continuously evolving biosphere.

More subtly, the slowly growing continents must have directly changed the Earth's reflectivity since land is more reflective than sea. Furthermore, the land's reflectivity altered as it became colonised, first by bacteria, then by lichens, plants and, finally, animals. The position of the land also changed through time as a result of continental drift and this affected the reflectivity through its influence on ice build-up. The world tends to be colder during times, such as today, when there is a large continent at one of the poles onto which a thick ice cap can grow. The position of land also affects ocean currents, and if these are blocked from moving heat from the equator towards the poles, then once again there is a build-up of ice at the poles leading to an increase in reflectivity and a drop in the world's average temperature. These changes in reflectivity due to varying clouds, varying ice cover, changing land-mass size and changing continental cover could theoretically have caused the Earth's temperature to alter by as much as 80°C. Add that to the effects of the variation in solar heat output and the potential changes in temperature could have been as much as 100°C.

These calculations ignore the effects of changing greenhouse gases, and these are affected by geological processes as well as by biological ones. For example, when mountain building is intense, weathering also increases and this takes carbon dioxide out of the atmosphere. The rate at which organic carbon is buried is also affected by geological factors such as the location of the continents and the depth of the oceans, and this also causes carbon dioxide levels to fluctuate through time. If we include such greenhouse gas variations, the potential climatic variability of our world becomes enormous. Given this, it really shouldn't be surprising that the Proterozoic and Phanerozoic experienced fluctuations in temperature. The real surprise is that the temperature changes were so small. Solar luminosity, planetary reflectivity and atmospheric composition have all changed dramatically during the Earth's 4.5 billion-year history and, in principle, these factors could have conspired to create very inhospitable temperatures on our planet. A completely ice-covered Earth, illuminated by a faint young Sun, and with no greenhouse effect, would have had an average temperature of 90°C below freezing. On the other hand, if the Earth had similar properties to Venus and was warmed by our present Sun, it would have warmed to over 400°C.

However, none of these massive potential changes in temperature ever actually occurred. The coldest our world has been was during snowball Earth episodes when globally averaged temperatures were certainly below freezing and may have been as low as −50°C. At the other extreme it is unlikely that temperatures over the last 4 billion years were ever more than 60°C for long, since this would have led to a runaway greenhouse effect and the Earth would now resemble Venus. Temperatures may have changed by tens of degrees over the billions of years of Earth history but certainly not by the hundreds of degrees of fluctuation that could have occurred. The Earth has therefore had surprisingly stable temperatures for a very long time and the various geological, biological and astronomical influences on climate must therefore have somehow nearly cancelled each other out.

There are three possible explanations for why we happen to live on such a well regulated planet and, with a little poetic licence, these can be characterised as God, Gaia or Goldilocks. God seems like a very straightforward explanation; we live on a planet with good weather because God made it that way. However, I don't believe that we should resort to transcendental explanations for what we see in the physical Universe around us. History shows science to be better equipped than religion when seeking that particular kind of enlightenment. If you're looking for God, I wouldn't start from here.

What of Gaia? Gaia has already been alluded to in this book, but to recap: the Gaia hypothesis postulates that life itself regulates our planet's climate in a way that ensures the biosphere thrives. Gaian proposals can be thought of as an extreme form of the conventional scientific view discussed earlier that says that climate stability is the result of natural geological and biological feedback mechanisms that automatically moderate climate change. As I mentioned before, this hypothesis may have confused cause and effect but biological feedback control does need to be taken very seriously as an explanation for the continuous good weather our planet has enjoyed.

The final explanation, Goldilocks, is that, just as with the third bowl of porridge from *Goldilocks and the Three Bears*, the temperature is 'just right' purely by good fortune. That, of course, is what this book is about. Perhaps planets where frequent, severe climate catastrophes happen are much more common than worlds with long periods between disasters. Perhaps planets that cool too quickly, or too slowly, are also far more common than planets like the Earth where the effects of solar and Earth evolution happen to roughly cancel each other out. Intelligent observers couldn't look out onto such worlds because observers would be unlikely to evolve in the impoverished biospheres that would result. Again, we can't conclude anything about what is normal or natural when considering properties of our planet that were essential preconditions for our existence.

Distinguishing between Gaia and Goldilocks is a major theme in this book. I will come back to Gaia later for a more detailed look before suggesting ways we might tell Gaia and Goldilocks apart. But first, I want to look at one final climate-related topic: the Earth's recent ice ages.

8
Music of the Spheres

The stars and planets were much more familiar to our ancestors than they are to most people today. The absence of bright street lights, until very recently, partly accounts for this greater familiarity but it was also due to the fact that their survival as farmers, hunters or gatherers relied on understanding the passing of the seasons as they are reflected in the changing constellations visible through the year. Simple observations of the night sky told farmers when to plant or harvest their crops and told hunter-gatherers when to look for migrating prey. One of the best known examples of this comes from ancient Egypt, where the first appearance of the brightest star in the sky, Sirius, heralded the annual flooding of the Nile and was anxiously awaited every year. Thus, regularities in the heavens were both useful and reassuring. Given this background, it's not surprising that all early civilisations were confused and intrigued by the planets (from the ancient Greek word *planetoi* meaning wanderers): their positions change, making the timing of their reappearance each year much harder to predict than that of the fixed stars. It seemed reasonable to assume that planets, like the fixed stars, foretold something useful but that the gods had made the message more cryptic and readable only by an elite priesthood. Thus was astrology born.

Astrologers have therefore long held that planets affect our destiny. They are, of course, completely correct. Gradual changes in planets' orbits have climatic effects and, if this book has a theme, it is that climate is destiny (or at least a precondition for destiny). The orientations and shapes of planetary orbits slowly oscillate over tens of thousands of years and these mild orbital vibrations, along with moderate wobbling of the Earth's axis, drive the ebb and flow of ice ages. My tour of past climate change is therefore incomplete until I have described the gentle celestial dance of our solar system and how this has driven glaciations, especially over the last 2.5 million years, during which we have been living through one of the Earth's rare ice ages.

The discovery of the ice age, repeated episodes of significantly colder climate in the geologically recent past, is one of the great achievements of 19th-century science. The key moment was a meeting of the Swiss Natural History Society in the town of Neuchâtel in 1837. Neuchâtel had been chosen to host this prestigious meeting because of the irreproachable reputation of the curator of the town's museum, the biologist Louis Agassiz. Expectations were high that, in his welcoming speech, Agassiz would reveal some of his latest findings on fossil fish; work for which he had rightly garnered international respect. Instead, Agassiz described evidence that Switzerland had once been the location of a massive ice cap to rival that of modern Greenland. His audience was shocked, and none more so than the people who had shown him the evidence, his friends Jean de Charpentier and Ignatz Venetz. They'd only been trying to convince Agassiz that Switzerland had once been marginally cooler; Agassiz's reinterpretation that vast swathes of the entire world had been covered in ice took things much further than they were ready to believe. Agassiz's proposed ice age seemed a regressive step at a time when the young science of geology was still struggling to establish that the rocks of the Earth showed the cumulative working of natural processes over immense periods of time rather than the effects of biblical catastrophism.

Agassiz therefore needed to work hard to strengthen support for his ideas, and he visited Scotland in 1840 specifically looking for evidence that a country that now has no glaciers at all had been the centre of vast ice sheets in the past. He found what he was looking for and it convinced him that much of the northern hemisphere had once been covered by thick ice extending from the pole down to the latitude of the Mediterranean. The evidence was there for all to see. Moraines, for example, chaotic piles of boulders and sediment bulldozed into ridges typically tens of metres high by glaciers, were soon found in Scotland by Agassiz and others. Other features too can be found throughout northern Europe, such as erratic boulders the size of houses sitting in fields after being carried by ice many kilometres from their place of origin. The Scottish lochs and Norwegian fjords are also signatures of recent glaciation: straight, steep-sided channels are characteristic of valleys carved by ice. These valleys became flooded after the glaciers retreated, forming the characteristic long, narrow water bodies. Within a few decades, unmistakable glacial hallmarks were also found in other parts of Europe and in North America and Asia; these showed conclusively that Agassiz's idea of a widespread ice age was accurate.

In fact, there have been many such ice ages or, more accurately, many phases of glaciation within a single ice age that has unfolded over the last 2.5 million years. Agriculture, civilisation and the whole of written human history have all occurred since the current 'interglacial' period of warmer weather began just over 11,000 years ago. We can't be sure when this brief interval of more clement weather will end, but interglacials tend to be much shorter than the periods of truly cold climate that they punctuate. What drives this ebb and flow of ice caps? The idea that it was due to changes in the Earth's orbit was suggested almost immediately, although this remained a controversial idea until very recently.

Mankind's quest to understand planetary motions probably began when we first noticed these wanderers in the sky, perhaps 200,000 or more years ago. We can only speculate about what early watchers of the sky thought of the complex movements of the planets, but a desire to find an underlying beauty and simplicity is clear by the time of the semi-legendary Pythagoras who was born on the Greek island of Samos in about 570 BC. Pythagoras is best remembered today for his eponymous theorem; the one about the 'square on the hypotenuse' that links the side-lengths of a right-angled triangle. This early philosopher is a rather shadowy figure. We can't be sure which achievements were truly his own and which came from his followers and successors. Even his theorem may have been discovered centuries earlier by Egyptian, Indian or Babylonian mathematicians, although this suggestion is controversial and unproven. Nevertheless, even if Pythagoras and his followers didn't discover it themselves, 'the theorem of Pythagoras' was just what they were looking for, because they believed that links between geometry, numbers and the natural world were spiritually significant. In a search for similar relationships to bolster these beliefs, the Pythagoreans stumbled on another equally interesting connection, one between geometry and musical harmony. They discovered that two strings, identical in every way except for one being twice the length of the other, produce notes separated by exactly one octave. More than this, they found that all harmonious pairs of musical notes have string lengths with simple ratios. Strings with lengths in the ratio of two to three, for example, are aurally separated by what musicians call a 'fifth' and they sound pleasing together. This discovery reinforced the Pythagoreans' belief in the interconnectedness of all nature and encouraged them to look for similar relationships in the heavens. They began to search for a 'music of the spheres': harmonious relationships linking the planets to one another in a similar way to the links they had found between harmonious notes.

Two thousand years later this rather vague idea was taken up by Johannes Kepler, a German mathematician born in 1572, who was much more precise about what exactly celestial harmony should look like. Kepler was an early supporter of Copernicus' heliocentric ideas. He appreciated the simplicity of Copernicus' Sun-centred scheme, compared to the complex apparatus needed to explain planetary motions if the Earth lay at the centre of the Universe, and he wanted to reinforce this elegance. More specifically, he wanted to understand the architecture of the solar system; the way that the worlds are distributed through space. It was already known that the planets are crammed relatively closely together in the inner solar system but are more spaced out as we move away from the Sun. Kepler believed that this pattern was important and that the increasing separations of successive worlds should obey some simple but mathematically beautiful rule. Along with most astronomers of this pre-Newtonian time, he imagined each planet to be embedded in the surface of a crystal sphere centred on the Sun, so that the solar system consisted of a set of nested spheres. He then suggested that five very special shapes, the Platonic solids, could be inserted precisely into the five gaps between these six spheres.

Platonic solids are a set of particularly simple three-dimensional shapes that were first investigated by the ancient Greeks. The best known of these is the cube, a shape made from six identical square faces. The tetrahedron, sometimes known as the triangular pyramid, is another shape that can be made from identical faces; in this case four equilateral triangles. All the Platonic solids have this property of being constructed from identical faces that, in turn, have sides of identical lengths. The Greeks discovered that there are only five such shapes, the other three being the eight-faced octahedron, twelve-faced dodecahedron and twenty-faced icosahedron. Kepler was impressed that the number of gaps between the six known planets was exactly equal to the number of Platonic solids and so he wanted this to be a law of Nature – an expression of the world's beauty and harmony. In Kepler's scheme a cube, for example, just fits inside Saturn's sphere while perfectly enclosing the sphere of Jupiter. Other pairs of orbits are similarly filled by the remaining Platonic solids.

This recipe for constructing a solar system sounds hopelessly mystical to the modern mind and turned out to be completely wrong. For a start, this solution to the Pythagorean search for celestial harmony doesn't even give the right architecture for the solar system. A cube between the spheres of Jupiter and Saturn places Saturn 86 million kilometres closer to the Sun than it really is. This became obvious within Kepler's lifetime, but even if the theory had predicted orbital sizes with sufficient accuracy to be more credible, it would have been dealt a mortal blow by the discovery of Uranus in 1781. There were no Platonic solids left to place between Saturn and this new world. However, enlargement of the solar system's retinue of planets still lay almost 200 years into the future and so Kepler initially persisted with his ideas. This was one of the most productive mistakes in the history of science. While trying to prove a link between Platonic solids and planetary spacing, Kepler stumbled instead on the first mathematically accurate descriptions of planetary orbits: Kepler's laws of planetary motion. Kepler's first law simply states that orbits are not circular, as everyone had believed, but are instead elliptical. His second law describes how the orbital speed of planets increases and decreases as their elliptical orbits bring them closer to, or further from, the Sun. Finally, in 1619 and ten years after publication of the first two laws, Kepler revealed a third relationship that showed how the duration of an orbit increases with distance from the Sun. This third law explains why Mars's year is twice as long as our own, for example.

Kepler's laws are still used today and they were among the first-ever examples of mathematical formulas that accurately describe a facet of our Universe. If Kepler hadn't discovered them it is debatable whether the mathematical laws of mechanics would have been formulated by Galileo in the 1630s and whether Isaac Newton's law of gravity could have been published in 1687. Kepler's somewhat misguided search for celestial harmony was therefore a fruitful one since it played a major role in the emergence of modern science. In any case, it should be seen as part of a long tradition that continues to this day of looking for beauty in our descriptions of nature.

The Pythagoreans' search for a link between musical harmony and solar system architecture therefore ultimately failed, but another aspect of how strings vibrate does turn out to be useful for understanding planetary systems. The Pythagoreans investigated *the* note produced by a string but, in reality, no string produces a pure note. Instead, the dominant pitch of a string is contaminated by fainter contributions from other notes, overtones as they are known, and they give the sound of a string its unique character. In exactly the same way, planetary orbits do not wobble at a single frequency, but instead undergo a combination of many different vibrations.

To explain this more clearly, I'll stick with musical character for a little while longer before coming back to planets. All musical instruments produce notes that include overtones but it's the different combinations of these that give them their distinctively different sounds. If all instruments produced a perfectly pure note they would sound identical and music would be far less beautiful. Stringed instruments, trumpets and glockenspiels vibrate in unique ways that combine many tones, and so they sound quite different to one another. Any instrument playing a middle C will vibrate about 260 times per second, but superimposed upon this fundamental note will be fainter vibrations at other frequencies. For example, most instruments also vibrate at twice this rate, 520 times per second, but the exact size of this additional contribution to the sound will vary from one case to another. Brass instruments do not produce very much of this so-called first harmonic and, instead, vibrate strongly three times faster than the fundamental note at 780 times per second. It's the near-absence of every other harmonic that gives that distinctive 'brassy' sound. A piano string, on the other hand, produces a gradual dropping off in intensity with each increasing harmonic. Overtones of many instruments also include vibrations at frequencies that are not simple multiples of the fundamental note, and with these too the exact contribution to the overall sound varies from one instrument to another.

This idea that the characteristic sound of an instrument results from adding together a fundamental note and an instrument-dependent set of overtones applies equally well to the behaviour of planets. The planets in our solar system have orbits that vibrate at a fundamental period with oscillations at other periods, overtones if you like, added in to give a unique character to the wobbling of each world's orbit. Imagine each orbit as a hoop that, instead of lying flat, has been tilted a little. The direction and size of the tilt slowly changes through time, and each hoop's shape also oscillates between being slightly oval and being circular. Planetary orbits, unlike musical instruments, therefore sing two songs simultaneously: one is the song of orbital orientation and the other is the song of orbital shape. However, there is also an important similarity. The solar system is like an orchestra tuning up before a performance. The individual instruments, the planetary orbits, all play the same notes but they exhibit distinct combinations of the overtones so that each planet, like each instrument in an orchestra pit, sounds a little different.

These orbital vibrations are generated by the gravitational attraction between the planets. A simple solar system consisting of a single planet orbiting the Sun is rather boring. The planet goes round and round the Sun for ever in an elliptical orbit that never changes its shape, never changes its orientation and never does anything interesting at all. The picture gets a little more complex if the planet, or the Sun, is not a perfect sphere but I'll wait until later to describe some of the trouble that difficulty causes. For now, I want to look at a rather different complication. In the late 18th and early 19th centuries the French genius Pierre-Simon Laplace and the equally brilliant Italian, Joseph-Louis Lagrange, accurately showed for the first time how adding more worlds to a one-planet system made the behaviour of the system much more interesting. The orbits wobbled and changed their shapes over tens and hundreds of thousands of years to generate a genuine 'music of the spheres' in which every planet sang the same notes but in their own distinctive voices.

Of course, we can't really hear the planetary orbits singing, because sound does not travel through the vacuum of space – and in any case the fundamental vibration of the solar system, a change in the orbital tilts over a period of 50,000 years, produces a note 50 octaves below middle C and far too deep to be detectable by human ears. Nevertheless, it would be fascinating to hear Jupiter's hum turned to sound and speeded up 400 trillion times to upscale the fundamental frequency into a middle C. We'd start with silence when Jupiter was alone, but bringing in Saturn would generate the expected pure note near middle C as the orientation of Jupiter's orbit started to wobble. We'd also hear that second song, another deeper note around F sharp, generated by the vibrating shape of Jupiter's orbit. Then, as we added, say, Uranus with its particular mass and orbital radius, new overtones would be brought to the songs and the fundamental notes would shift very slightly. The richness of Jupiter's song would increase further as each additional world was placed in its orbit until Jupiter was generating a complex hum. We could also listen to the songs of the other planets and there would be a family resemblance between them, because they would all be playing the same fundamental notes, but the different combinations of available overtones would create distinctly different songs.

Like going from a pure middle C to the sound of a violin string, adding planets generates beauty through complex over-tones, overtones that ultimately control the rhythms of our ice ages. The full family of eight planets produces oscillations with periods ranging from 46,000 years up to 4 million years. Many of these control changes in the orientations of the planetary orbits. The Earth's orbital wobbles, for example, are dominated by four or five vibrations the most important of which produces a 69,000-year variation in the tilt and tilt-direction of her orbit. Other overtones control the rate at which orbits change their shapes. The resulting main period of oscillation for the Earth's orbital ellipticity is about 400,000 years with a number of other, slightly less important, periods clustering near to 100,000 years.

So Pythagoras and Kepler were, in a sense, correct after all. The solar system does behave like a musical instrument and its notes are controlled by the size of the orbits, together with the planetary masses, in a way reminiscent of how string pitches are controlled by their lengths and tensions. There really is a music of the spheres but it emerges only as a result of a complex analysis using 18th-century mathematical tools undreamt of 200 years earlier. Indeed, the full complexity is only now being unravelled, as a result of 21st-century computer power.

There is yet another set of wobbles in the solar system that need to be included if we are to understand how orbital oscilla-tions affect ice ages. These occur because planets are not perfect spheres. All spinning worlds bulge a little around their equa-tors as centrifugal forces generate the planetary equivalent of middle-aged spread. The Earth, for example, is 43 kilometres fatter when measured across the equator than it is between the poles. These equatorial bulges cause the individual planets to sing solo, by wobbling on their own spin axes, at the same time as they participate in the collective hum of the solar system. The planetary, as opposed to orbital, wobble is exactly the same as that of a child's spinning top. I am watching a 'Thomas the

Tank Engine' spinning top wobble in front of me on my desk right now as I write this paragraph. The toy is spinning round several times a second, and as it does so, its handle is slowly drawing a small circle in space every few seconds. The gradually slowing spinning top is precessing, as this motion is called, more and more rapidly and the wobble has just become so violent that the edge of the toy has touched the desk and stopped its movement. Planetary rotation axes precess in an identical manner to my spinning top but in the case of the Earth it takes thousands of years, rather than seconds, for each wobble. Like the handle of the spinning top, the Earth's North Pole draws a giant circle in the sky, returning to its original position only after 26,000 years. At present, anyone standing at the North Pole and looking straight up on a clear night will stare almost directly at the star Alpha Ursae Minoris, better known as the pole star. This is a temporary, if convenient, fact. Precession of the Earth's axis slowly alters the orientation of our planet so that, a few thousand years into the past or the future, the North Pole no longer points at the pole star.

In the case of the spinning top, precession is caused by gravity attempting to pull the toy over. However, it's not easy to change the orientation of a spinning object, which is why spinning tops and motorbikes generally stay upright (bicycles are more complicated as there's an additional effect involved). On the other hand, it is relatively easy to make the axis of a spinning object precess. If you get the chance, try it with a bicycle wheel. Hold the axle between your two hands and get a friend to spin the wheel. Then try drawing a circle in the air with the axle. You'll find you can do it if you move the axle in the same direction as the wheel rotation, but that it is almost impossible if you try to go the opposite way. You'll also find it very difficult to move one end of the axle in a straight line without it wobbling in a circle instead. So, the spinning top is trying to fall over but, because of gyroscopic effects, it is unable to do so and precesses instead.

Slow wobbling of a planet's spin axis in space is caused in a similar way. Tidal forces tug on the equatorial bulges in an attempt to bring planets into 'more upright' positions. The Earth's axis, for example, is not perpendicular to its orbit but is tilted by about 23 degrees. This angle is called the obliquity, and tidal forces try to reduce its size as they tug on the equatorial bulges. Solar-generated tides occur because the day-lit side of a planet is very slightly closer to the Sun than the night side. Gravitational attraction by the Sun is therefore marginally stronger on the day side, and this difference in gravity across the planet produces the stretching force responsible for tides. Additional tidal stresses are produced by moons when a planet also possesses these. These forces try to reduce a planet's obliquity as they tug on equatorial bulges, but the tussle between tides and gyroscopic stability is a fight that neither side wins; precession of the planet's axis occurs instead. As you might expect from all this, precession speed is controlled by the size of the tidal forces, the rate of spin, the obliquity and the size of the equatorial bulges.

To recap, the picture I have tried to draw here is of a dynamic, almost vibrant, solar system in which the orbits of the planets are constantly changing their orientation and shape while, at the same time, the axes of the planets themselves are experiencing continuous variations in the directions they point in space. Fortunately for us these astronomical oscillations are much smaller in the solar system than they could be but, nevertheless, this music of the spheres is loud enough to affect our climate. In particular, these celestial rhythms are undoubtedly the main factors controlling the ebb and flow of Earth's ice ages.

Lagrange's work on the solar system's oscillations was already 40 years old when Agassiz dropped his bombshell about the existence of ice ages and so an astronomical driver for the growth and decay of the glaciations through time seemed an obvious explanation. However, as I've tried to emphasise above, the song of the solar system is quietly sung and its climatic effects should be tiny. At most, the heat received at any given season and place on the Earth's surface changes by just a few per cent, and averaged over an entire year there are no significant changes in heating at all. Astronomical driving of the ice ages was therefore rejected until an extraordinary Glaswegian entered the scene.

In 1859 the 40 year-old James Croll became caretaker at a school in Glasgow after previous careers as an insurance salesman, millwright, carpenter and farm labourer. At Anderson College this underprivileged but highly intelligent man finally fell on his feet, because the college had an extensive library that Croll was allowed to use. He devoured its contents. Croll developed wide interests in scientific, theological and philosophical matters but he decided that, in particular, he wanted to solve the problem of the cause of the ice ages, and by 1864 he was publishing papers that drew the attention of the scientific establishment. Croll believed that build-up of ice sheets was the key to the problem. He suggested that, when the Earth's orbit was particularly elliptical, the Earth would spend part of the year significantly further from the Sun, and if this corresponded with winter in one of the hemispheres, ice sheets would grow more extensively during the resultant severe winters and would not melt back completely in the summer. Glaciers would therefore get bigger and cool the planet. Hence, Croll was an early proponent of the ice-albedo positive feedback mechanism that I described in an earlier chapter. Ice-albedo feedback magnifies the climatic effects of orbital changes so that very slight changes in how the Earth is heated lead to relatively large changes in temperature. To test this idea, Croll used Laplace and Lagrange's theories to calculate how ellipticity has changed through time. This is an extraordinary accomplishment in itself. I've done similar calculations using spreadsheets, and with the help of these, the hundreds of calculations necessary can be

completed in a few minutes, but Croll had to do all the arith-
metic by hand. This tedious and error-prone work must have
taken him months. But what he found was encouraging. The
best guesses of geologists for the timing of the cold periods
coincided with Croll's calculations for the times of greatest
orbital ellipticity.

As a direct result of the impact he made through publishing this work, Croll was given a job at the Scottish Geological
Survey in 1867, and by 1875 he had published a book setting
out all his ideas in a single place. Croll's astronomical theory of
ice ages seemed triumphant, but within a few decades problems
began to emerge. For a start, the theory implied that ice ages
alternated between hemispheres. Sometimes the northern hemisphere would experience an ice age; at other times, the southern
hemisphere would. Unfortunately, evidence was emerging of
simultaneous cooling in both hemispheres. The geological dating of glaciations was also being revised and the newer results
no longer agreed with Croll's calculations. The astronomical
theory fell out of favour.

However, the fortunes of the astronomical theory of ice
ages were as cyclic as the phenomena it set out to explain, and
belief in this mechanism came back into fashion again with
publication, in 1920, of a paper by Milutin Milankovitch, a
Croatian/Serbian professor of applied mathematics who had
built his reputation analysing properties of the new wonder
building material: reinforced concrete. Milankovitch's ice age
work was summarised in his book of 1941 in which he took
Croll's calculations a stage further and used knowledge of the
orbital oscillations to calculate the expected changes in temperature at different seasons and different latitudes. What he
discovered was that known glaciations corresponded to times of
cooler summers at moderately high northern latitudes. This was
almost the opposite of Croll's assumption. It wasn't cold winters
but cold summers that were important, and this was because

they reduced the melt-back of glaciers in the summer months. Milankovitch's results showed that the Earth becomes colder when summers fail to melt the snow rather than when winters generate more ice. The fit of this new theory to geological data was impressive but, unfortunately, once again misleading. Subsequent, more accurate, dating of glaciations together with the realisation that some of the key deposits were not glacial at all destroyed the beautiful correlation between Milankovitch's theory and the geological evidence. Scepticism over the astronomical theory grew once again.

The breakthrough finally came in 1976 with publication of a classic paper by James Hays, John Imbrie and Nicholas Shackleton relating astronomical drivers to climate data obtained by drilling in the ocean floor. Analysing sediments on the deep ocean floor has the great advantage that this gives almost perfectly uniform and continuous data extending back hundreds of thousands to millions of years. There has been an internationally funded programme of scientific drilling into the sea floor from 1968 through to the present day and the IODP (Integrated Ocean Drilling Programme), as it is now known, continues to provide some of the best scientific data on the history of our planet available from any source. The cores from this programme can be used to provide climate data such as oxygen isotope analysis (which indicates ice volumes in polar regions, as discussed earlier) along with analyses of planktonic remains that can give an additional indication of the sea surface temperature and salinity at the drilling site. By taking cores from a range of depths below the sea floor it is possible to see how these factors have changed through time. The work of Hays, Imbrie and Shackleton, along with many subsequent studies by other scientists, produced a beautifully simple result. There is a clear, 41,000-year cyclicity in the climate data superimposed on other cycles at 100,000 years and at about 20,000 years. In more detail, the 100,000-year cycles dominate over the last

700,000 years with the 41,000-year cycle dominating before that. Periods of about 20,000, 40,000 and 100,000 years are exactly what the astronomical theory predicts. Let me start with the 41,000-year cycle that results from changes in the tilt of the Earth's axis.

Changes in this obliquity affect the intensity of our seasons. As a direct consequence of axial tilting, the northern hemisphere gets increased sunlight between April and September, which gives us summer, and reduced sunlight between October and March to produce winter. Of course, when the northern hemisphere is tilted away from the Sun, the southern hemisphere is tilted towards it and so the southern seasons are the reverse of those in the north. As the Earth's axis and its orbit both wobble in space, the angle of obliquity changes by a degree or two and that alters the intensity of the seasons. When obliquity is small and seasons less intense, there is less melting of the polar ice caps in summer, making the Earth more reflective and cooler. Changes in the Earth's obliquity occur more slowly than the 26,000-year axis precession because of the additional 69,000-year cycling of the Earth's orbit itself and it's the combination of these two wobbles that produces a changing obliquity. The resulting 41,000-year oscillation produces the signal so clearly seen in the ocean sediment data.

Axial precession of the Earth has another climatic effect that results because the Earth's orbit is not quite a perfect circle. Today we are closest to the Sun in December when the northern hemisphere of the Earth is tilted away and experiencing winter. Conversely, we are furthest from the Sun in the northern summer. This mismatch between season and distance has the effect of reducing the seasonal intensity in the northern hemisphere. However, the southern hemisphere experiences exactly the opposite effect, since we are close to the Sun during the southern summer and furthest in the southern winter. Thus Antarctica, where the bulk of the world's ice resides, is currently experiencing relatively intense summers that melt back much of the ice cap. However, precession of the Earth's axis changes this relationship slowly over time. The timescale is again complicated by the fact that the Earth's orbit is also precessing but the effect is to change the relative intensity of southern versus northern seasons on a timescale of about 20,000 years. This 'climatic precession' effect produces the 20,000-year cycles seen in the sea floor data.

Finally, the 100,000-year fluctuations in climate seem to be directly related to changes in the eccentricity of the Earth's orbit, almost exactly as predicted by Croll except that cooling occurs when eccentricity is low rather than high. The reason is, as stated earlier, that glaciers grow when summers are cool rather than when winters are severe. It remains a puzzle why the effect of an elliptical orbit is so strong when the corresponding changes in seasonal intensity are relatively small. It is also unclear why there was a flip from dominance by 41,000-year cycles to dominance by 100,000-year cycles in the last million years. Despite these remaining uncertainties, astronomical forcing of the ice age fluctuations is now a well-established idea and rarely causes scientific controversy. But it is not the whole story.

Ice ages are actually rather rare in the Earth's history. We are, as I have said, living through one now, but to find an earlier ice age we have to go back almost 300 million years. There was yet another, probably milder, ice age around 450 million years ago but, prior to that, we have to go back to the Proterozoic snowball Earth episodes to see other occasions when the Earth was at least as cold as today. Through the majority of Earth's history our planet has been much warmer than today and almost completely free of any sea ice at all. Ice ages therefore clearly need some of the other factors I've previously discussed, such as the existence of a continent at the pole, blocking of marine currents or uplift of particularly large mountain ranges. At present we seem to have a triple-whammy of all three going on.

Given all this, you'd probably expect that climate is strongly affected by astronomical cycles only during these rare times of ice age conditions when ice-albedo feedback is able to amplify the small astronomical effects. At other times there is very little ice around, even in the winter, and so there is no opportunity for the ice-albedo effect to magnify the weak influence of the solar system's song. However, many, but certainly not all, geologists believe that Croll–Milankovitch cycles can be detected throughout the rock record, even at times of unusually warm climate. This may well be correct but it is hard to explain and remains a highly controversial area. Fortunately, it is not an important consideration from the viewpoint of this book.

Much more important is another qualification I should make concerning the contents of this chapter before we move on. I have, perhaps, exaggerated a little the harmonious nature of planetary motions. The solar system's 'music' is actually a rather inharmonious sound; more seagull than nightingale. None of the solar system's notes would actually sound pretty together because they lack the simple relationships that Pythagoras found for harmonious strings. We'll see later that this is fortunate. True harmonies in the heavens would lead to catastrophe because the normal, almost clockwork, regularities in the motions of the planets would break down and give way to unpredictable chaos. This is a fate we barely avoided! However, that's for later. First, I'd like to move from geology and astronomy to biology. It's time now to consider the possibility that life, itself, is the main guarantor of its own survival.

9

Force of Nature

Is biological evolution alone powerful enough to explain life's 4 billion years of success on an ever-changing world? The great diversity of organisms that share our planet certainly suggests that life is extraordinarily adaptable. Many people have been so impressed by this diversity and adaptability that they imagine life to be capable of colonising almost any environment eventually. This chapter will look at the power that nature has shown to adapt to almost everything the Earth has thrown at it. However, as we'll also see, Earth life does have limits, and these may be sufficiently severe and universal to restrict living creatures, particularly more complex organisms, to conditions found only on the very best of worlds.

There is no doubt that life is remarkably versatile, as one recently discovered animal demonstrates particularly well. This creature lives at the bottom of the Mediterranean in pockets of salty water too dense to mix with the overlying sea. These sea floor brine pools completely lack oxygen and are largely inhabited by micro-organisms that do not need or want this gas. But these microbes do not have the stagnant, saline pools to themselves; animals have penetrated even this apparently hostile environment. In 2010, Italian and Danish scientists reported the discovery of tiny creatures that spend their entire lifetimes in these pools and never, ever breathe oxygen. The animals are about a third of a millimetre long and look a bit like miniature jellyfish, although this is misleading since they are a type of *Loricifera*, a group of animals much more closely related to insects and crustaceans, or even us, than jellyfish.

Remarkably, this particular species of *Loricifera* gets its energy by combining carbohydrates with hydrogen sulphide produced by bacteria in these putrid brine pools. It would be hard to conceive of a better illustration of the extraordinary power of evolution than an animal that doesn't need to breathe oxygen, and given examples like this, it is entirely understandable that many observers believe life will thrive under almost any conditions. Nevertheless, there are limits to life's ability to cope with difficult conditions. To see why, we need to look at evolution in a bit more depth.

Modern evolutionary theory begins in 1859 with Darwin's publication of *On the Origin of Species* following his receipt of the letter from Alfred Russel Wallace mentioned in Chapter 2. However, the idea that species have gradually changed over time, that they have evolved, was widely accepted well before then. The fossil evidence that ancient animals and plants were different from modern ones was overwhelming and the case was strengthened even further by centuries of thorough anatomy showing clear family resemblances under the skins of many species. The reality of evolution was plain for all to see, and many people had discussed it in the late 18th and early 19th centuries. Darwin's contribution was to propose a convincing mechanism for evolution and his explanation has all the elegance and inevitability of a well-crafted mathematical theorem. He noted that offspring are similar to, but slightly different from, their parents and siblings. Hence, some children are better suited than their brothers and sisters to the particular conditions – of climate, food availability, predation and so on – that they find themselves living in. These better-adapted offspring are more likely to survive to produce children of their own, children who will probably share their parents' beneficial characteristics. As this process is repeated over many generations the fraction of the population owning these desirable traits steadily grows, and so the entire species gradually becomes better adapted to its environment. This is evolution by natural selection: a mechanism powerful enough to turn fish into giraffes, given 400 million years, and a mechanism powerful enough to explain all the fossil, anatomical and biochemical evidence that makes evolution an undeniable fact of life.

Evolution by natural selection therefore clearly explains much that we see in nature; but, since it can be expressed as 'survival of the fittest to survive', it has been criticised by some as a mere tautology. It seems to me that expressing Darwin's idea this way simply emphasises how inescapable the process is. Natural selection is an unavoidable consequence of the struggle for survival and I sometimes wonder what it is that sceptics think stops evolution from happening. One suggestion that they might make is that evolutionary change should become imperceptible once organisms approach optimal adaptation. However, this is prevented by the fact that the environment also changes through time. As this book emphasises, the Earth's climate and geography have varied throughout our planet's history and, even more importantly, the living environment itself has changed as new species emerged and old ones died out. Under these conditions natural selection became an arduous run on a treadmill rather than a quick sprint towards a finishing line. And this was a treadmill that ensured continuous fitness in a constantly changing world. The ability of natural selection to tailor living things to fit almost any environment created by our dynamic planet is certainly impressive, and the anoxic animals mentioned above are a stunning illustration of that power. However, Darwin's mechanism is not quite omnipotent. Evolution, like politics, is the art of the possible: both are constrained by laws and by the baggage of history.

Let's start with a look at how the laws of physics and chemistry restrict what is possible even in principle. There must, for example, be a minimum and a maximum temperature for life. I don't believe that I am being naive or anthropocentric to suggest that even the most alien imaginable biology will have temperature limits. At very low temperatures, say $-270°C$, chemical processes are far too slow for any conceivable form of metabolism to occur even on multi-billion-year timescales. At temperatures of tens of thousands of degrees centigrade on the other hand, all atoms and molecules are destroyed to leave only a structureless soup of electrons and nuclei and, again, no interesting metabolism can occur. However, these are very extreme limits that do not place severe constraints on planet-based life-forms. Are there reasons why the life-friendly temperature range might be significantly narrower than this? Could the range even be small compared to typical temperature ranges on typical planets? Absolutely, and you only need to do a bit of cookery to see why.

Cooking is fun, but that's not the only reason it is a universal human activity. Obviously heating food makes it taste better, unless I'm the cook, but the more fundamental reasons for cooking are that it kills parasites and that it chemically alters the proteins and carbohydrates from which food is made. The benefits of killing parasitic organisms before we eat them are obvious, but the chemical reactions are equally important because they produce interesting new flavours and frequently make inedible, or even toxic, substances palatable. As a result, cooking has greatly increased the range of foodstuffs available to human beings and it is fair to say that cookery is one of our greatest inventions. The chemistry of cooking is closely linked to its parasite-killing ability; altering the chemical make-up of a living being by heating it is usually fatal, as lobsters discover the hard way. One reason for this is that proteins work only over a limited temperature range. Proteins are complex, long-chain molecules that get their remarkable life-enabling properties from the very specific shapes they form when folded up. Haemoglobin, the protein in blood that transports oxygen, is a good example of the importance of folding because its shape, and changes in its shape, allow it to both grab and release oxygen molecules. High temperature disrupts the function of proteins because it unfolds them. Furthermore, once unfolded by heating, most proteins do not refold into their correct, biologically useful form when re-cooled.

Despite this, some micro-organisms can withstand the high temperatures encountered during cookery. Such heat-resistant microbes are found naturally in the hot springs of volcanically active regions on the Earth's surface and in similar settings, known as hydrothermal vents, found on the sea floor kilometres below the surface of the world's oceans. The record, so far, is a deep-sea microbe known to reproduce at a scorching 121°C and able to survive at up to 130°C. At these temperatures you might think these microbes are not in liquid water but must be living in steam; however, at great depths in the oceans, high pressure keeps water in liquid form. These thermophiles, as the heat-loving bugs are called, don't just tolerate the near-boiling conditions found in these places – they actually thrive in them and struggle to survive at what we would consider normal temperatures. Thermophiles have proteins that are relatively stiff and not easily unfolded, but this inflexibility makes it difficult for them to perform their normal functions at lower temperatures.

Hence, though organisms can adapt to life in hot water, the result is a creature unable to live at more normal temperatures. In contrast, other organisms have the much floppier proteins needed to thrive in extremely cold conditions. One example of such a psychrophile (as these cold-lovers are called) is *Chlamydomonas nivalis*, a species of algae responsible for 'watermelon snow' – pink, aromatic snowdrifts that sometimes appear on mountaintops and in polar regions. These microbes are, once again, restricted to a narrow temperature range and die if warmed even slightly. As a result of unfolding when too hot or excessive rigidity when too cold, proteins therefore work only over a limited temperature range regardless of whether they come from a thermophile, a psychrophile or a mesophile (a lover of moderate temperatures). Even the least fussy proteins, such as those taken from arctic fish that have to survive in a wide variety of temperatures, function only over a 20–30°C range.

Much of the chemistry of life on Earth is therefore very sensitive to temperature and this greatly limits the temperature range that organisms can tolerate. An extreme illustration of this occurs near the hydrothermal vents found on mid-ocean ridges, where as a result of upwelling magma creating new ocean crust, temperatures can vary from the usual, near freezing, values of the deep ocean to near boiling point over just a few metres. Even here, no organisms have evolved that thrive across a wide temperature range. One recent study of the deep sea vent shrimp *Rimicaris exoculata* by a French research group showed that it was dormant in the normal deep sea temperature of 2°C, suffered heat stress above 25°C and died in water temperatures much above 30°C. This is an organism whose main food supply, bacteria carried within its own gill-chamber, grows only in temperatures above 20°C. Vast clouds of these shrimps therefore lead a precarious existence hovering close, but not too close, to the hydrothermal vents that lie along the mid-Atlantic ridge. *R. exoculata* would gain a major survival advantage if it could better tolerate the occasional, but inevitable, accidental exposure to much warmer water, but toleration of such exposure has simply not evolved.

It seems therefore that life can withstand extreme temperatures but not extreme ranges. Of course, body temperature is not the same as environmental temperature and many species greatly improve their tolerance through adaptations such as internal heating, fur, feathers, shivering, sweating and panting. This increases the range of climates an organism can thrive in but, nevertheless, there is still a relatively small spread of temperatures over which any organism is comfortable.

The ability to survive extremes, but not extreme variation, also applies to environmental properties other than temperature. Microbes have evolved that can, for example, withstand high acidity, high alkalinity and high salinity. Organisms that thrive in severe conditions like these are called extremophiles and they are the subject of intense research. Partly, this research is directed at discovering the limits of life and whether bacteria exist that could survive on Mars or other planets. This research is also supported by the biotechnology industry, because proteins that withstand extreme conditions have proved useful as drugs and in diagnostic tests. The best known example of this is a genetic manipulation tool called PCR (polymerase chain reaction). PCR is used to create billions of copies of a DNA sample using polymerase (a protein that greatly speeds up DNA copying) extracted from the thermophile *Thermus aquaticus* which lives in hot springs. All living things make polymerase but the particular kind produced by *T. aquaticus* functions at temperatures high enough to break up DNA molecules. This combination of DNA copying and DNA fragmentation is the key to making PCR a relatively simple and quick process.

Extremophiles were discovered only in the last few decades but they have revolutionised the science of astrobiology. We now realise that life is much more robust than we had previously thought, and this improves the chances of its existing in what we would once have considered uninhabitable environments on other planets, such as the surface of Mars with its low temperatures and thin atmosphere that lets through intense radiation from the Sun. Some Earth organisms are able to cope well with cold, as I've already mentioned, and there are also microbes able to withstand Martian radiation levels. In the 1950s it was discovered that heavily irradiated, tinned meat can still go off and that the culprit was a previously unknown bacterium given the name *Deinococcus radiodurans*. This micro-organism has since been found growing happily in the most radioactive of surroundings such as the inside of nuclear power stations where it survives radiation levels thousands of times higher than those that kill people. *D. radiodurans* has been called the world's toughest bug but it may be about to lose this title: NASA has recently discovered two new microbes, currently with the rather dull names of strain-24 and strain-19, which are even more radiation resistant.

The existence of radiation-resistant microbes is rather surprising because high-radiation environments simply don't exist naturally on the Earth. Why have *D. radiodurans*, strain-24 and strain-19 evolved resistance to a non-existent danger? The answer almost certainly lies in the fact that they are also very resistant to desiccation; a threat to life that does occur in many Earth environments. These bugs are so tough that you can take nearly all the water out of them and, when you rehydrate them, they come back to life. To be able to survive severe drying, these bacteria have had to evolve efficient mechanisms for repairing their biomolecules because desiccation damages both genetic material and proteins. Radiation affects these biomolecules in a similar way and so mechanisms that repair cells after severe dehydration also repair damage sustained in highly radioactive environments. The ability of these extraordinarily tough little organisms to survive almost complete dehydration ironically illustrates the final, and most important, physical limit that I want to discuss in this chapter. All Earth life needs liquid water.

Despite the ability of organisms such as *D. radiodurans* to withstand severe dehydration, no known living things are biologically active in the absence of liquid water. *D. radiodurans* is not active when dehydrated; it is in a dormant state awaiting rebirth when water returns. Similarly, low-temperature organisms become dormant when their water freezes but also use anti-freeze to put this off as long as possible. Water in the liquid state is necessary because all of life's chemical reactions take place when compounds are brought together, either in solution or as a suspension, by water. Perhaps the best example of this is that one of life's most important molecules, DNA, can be copied only in the presence of water.

DNA is the molecule that stores the genetic information of all life-forms, and many viruses, on Earth. A copy of those genetic instructions is stored in nearly every cell and so, when single-celled organisms reproduce or when multi-celled organisms grow, additional copies of the instructions are needed. Thus, DNA copying is a central activity of Earth life. The ability to store genetic information and the ability to allow copying both result from the way DNA molecules are constructed. DNA consists of a long chain of chemicals that form a back-bone for the molecule. Each of the links in this chain has one of four molecules attached to it: adenine (A), guanine (G), thymine (T) or cytosine (C). The resulting string of bases (i.e. a string of As, Gs, Ts and Cs in a complex pattern) holds the genetic information that tells an organism's cells what proteins to make. Like computer memory, which stores digital information as a string of 1s and 0s, DNA molecules store genetic data as a string of As, Gs, Ts and Cs. Indeed, genetic engineers demonstrated in 1988 that man-made digital information can be stored in this way and, in the latest breakthrough in late 2012, American scientists have successfully written (and read back) an entire book, including pictures, stored in DNA.

The key to copying genetic information is that adenine will form a chemical bond with thymine while guanine will link to cytosine. As a result, DNA normally consists of two separate DNA molecules stuck together by a string of complementary bases. For example, the string ACTG on one strand will stick to the complementary strand TGAC on the other. In practice the two strands contain thousands of bases and the two long strands coil around each other to give DNA its famous double-helix structure. Copying a string of DNA is then a relatively simple procedure (I am, of course, ignoring lots of complicated details such as how the polymerase mentioned earlier in this chapter helps things along). If the two strands are gently pulled apart, in a mixture of bases and sugar-phosphate fragments suspended in water, each newly exposed strand will attach itself to complementary bases floating in the surrounding fluid and two new double-helixes will automatically form. Take the example from above in which ACTG was stuck to TGAC. If the first bases are pulled apart (i.e. A is separated from T) then the newly naked A will join to a T as soon as one floats by. If the next pair is then separated (i.e. C is now separated from G) then the C will pluck a G from the surrounding fluid. Sequentially unzipping the entire molecule results in a separated ACTG strand with a newly formed TGAC strand stuck to it. The original TGAC strand will, similarly, have a brand-new ACTG strand attached. Thus, where there was previously one DNA molecule (made of two strands) there are now two identical molecules (each made of two strands). The DNA has been copied. The new strands form efficiently because the bases are repelled by water and will try to hide themselves inside a DNA molecule as quickly as they can. So, as soon as a naked base appears as a result of unzipping of the existing DNA, a complementary base from the surrounding soup hungrily latches on and covers it up again. This process can't happen when water is absent and so copying does not occur in dehydrated DNA. There are no biochemical reactions more fundamental to life than DNA reproduction, and so Earth life must have liquid water to survive.

Liquid water might be essential but some organisms need surprisingly little of it. Hardly any rain or snow falls in the McMurdo Dry Valleys of Antarctica, and the little that does evaporates rapidly in the 'katabatic' wind of cold, dense air that streams ferociously down the valleys from the high Antarctic plateau at their heads. These valleys are among the driest places on Earth and, to make matters worse, temperatures average around −20°C and rarely exceed freezing point even in the height of summer. This combination of extreme cold and dryness is comparable to that expected on the surface of Mars and, as a result, the dry valleys of Antarctica are an area of intense research interest. Most organisms would find survival in the McMurdo valleys impossible, and yet this uncongenial location is not completely dead. The rocks are alive. Water percolates into the pore-spaces of the McMurdo boulders during rare snowfalls and this is much harder to evaporate than free-standing water. Rare, warm sunshine in the summer months melts this pore fluid, and so liquid water is at least occasionally available. Lichens and cyanobacteria are able to grow in the pore fluid just underneath the surface of the rocks, and other micro-organisms scrape a living within the mini-habitats these primary producers provide. This is one of the harshest environments on our planet but it is still successfully colonised by living organisms. Nevertheless, finding life in the McMurdo Dry Valleys is hard work and the contrast with, for example, the amount to be found in the almost equally cold but slightly wetter tundras of the Arctic is instructive. The deserts of our world are not biologically productive places.

Even if we try to be more open-minded and imagine very alien organisms, it is hard to envision how anything as complex as life could occur in the absence of a fluid medium for transporting chemical reactants. It is possible that other substances, such as ammonia which is liquid at temperatures below $-33°C$, could be used by alien life-forms but all compounds have a limited temperature range over which they are liquids. In fact, water has one of the greatest such spans and this is one of several reasons why it is particularly suitable as a solvent for life. It has been suggested that very dense gases could be used instead of liquids. However, these so-called super-critical fluids also have a narrow range of temperatures over which they allow interesting chemistry. Thus, whatever fluid is used, any conceivable form of biochemistry is likely to operate over only a limited range of temperatures.

So far I have discussed the limits to evolution imposed by the laws of physics and chemistry, but there are also limits imposed by the rules of the game of evolution itself, and these prevent organisms from always adapting to whatever is thrown at them by their environment. In particular, the rules of natural selection encourage cheating – adaptations that benefit individuals at the expense of the species. Cheating prevents evolution from producing truly optimised organisms, and the best known example of this involves the gender ratio, the proportion of males to females in a species. The majority of sexually reproducing species have very nearly the same number of males as females even though this is not usually the optimum ratio 'for the good of the species'. Take humans, for example. The human gender ratio varies from place to place and with time, but typically there are only about 5 per cent more boys born than girls. This slight excess reduces with age, because men have a slightly greater tendency to die from accidents and disease and so the gender ratio for breeding-age humans is close to 50–50. However, from the point of view of maximising the production of children, most men are surplus to requirements: a single man could father hundreds of children in a year while a single woman, with her much greater biological commitment to the process, could produce only one or two offspring in that time. A society in which there were a hundred women for every man could, in theory, produce nearly twice as many children per adult per year than would a society with a balanced gender ratio. So, why don't populations evolve towards a more optimal gender ratio in which females dominate?

The problem is that, in a hypothetical population that achieved a more sensible gender ratio biased towards females, men would be reproductively more successful than women since they could be fathering hundreds of children a year. Under these circumstances it pays to cheat; any mutant males or females who produce more boys than the species average will greatly increase the number of grandchildren they have. These numerous grand-children will probably share the reduced gender bias trait and so also be more successful than their competitors. The trait will therefore rapidly spread through the population and the fraction of boys born will steadily move towards a new, higher value with each successive generation. This story will repeat next time a boy-friendly mutation appears in the population and so, over evolutionary time, the gender ratio will move towards 50–50.

The tendency of natural selection to optimally adapt species to their environments can therefore be undermined by cheating, and this affects evolutionary processes relevant to this book such as adaptation to climate change. Natural selection can adapt organisms to cope with short-term fluctuations, such as those associated with the seasons, provided changes are frequent enough that an organism is likely to encounter them in its lifetime. Organisms with these adaptations will then thrive at the expense of those without them. Good examples are migration, where an animal simply moves away when conditions are poor, and hibernation, where organisms become dormant to get through tough times. Organisms can also cope with very long-term variations in climate, provided these occur gradually enough for the slow pace of natural selection to keep up. The problem comes when changes occur on an intermediate time-scale; significantly longer than a lifetime but shorter than the thousands of generations over which natural selection occurs. On these intermediate timescales, any attempt to adapt to the big picture by being a generalist tolerant of a wide range of conditions is thwarted by cheats: specialists that are well adapted to the specific conditions at any particular time. The specialists will kill off the generalists by out-competing them in the short term but will then die out themselves when conditions eventually change.

In the real world, biological responses to climate change are a lot more complicated than this. In mountainous areas biological adaptation may not be necessary, because plants and animals can simply move up and down the slopes to track climate changes. Yet other organisms may be pre-adapted to climate change by, for example, having a very broad geographical range requiring them to be tolerant or by being small, which, because of shorter generation-lengths and higher metabolic rates, often leads to faster natural selection. The response of organisms to climate change is therefore a complex, controversial and highly topical area of scientific research but, nevertheless, there does seem to be a connection between climate stability and biodiversity. There is, for example, evidence to suggest that the Earth's recent ice ages have substantially reduced biodiversity in the regions of the Earth most heavily affected, such as the northern areas of Europe and North America. It is also true that the most bio-diverse ecosystems on Earth, such as tropical rainforests and the deep ocean sea floor, are those with the most stable climates, although the exact reasons for this remain contentious.

There is one further way in which the inherent character of natural selection itself restricts the possibilities of what life can achieve. Each new generation is built on the last and so, as I indicated earlier in the chapter, evolution is held back by the baggage of its own history. If you'll forgive me, I'll try to explain this with a notoriously misogynistic quotation. For an 18th-century European, Samuel Johnson was relatively progressive in his views about women, but he still told his biographer that 'a woman's preaching is like a dog's walking on his hinder legs. It is not done well; but you are surprised to find it done at all.' I'm proud of being politically correct (it used to just be called good manners) but I couldn't resist including this well known quote because it's a great excuse for me to look at the subject of bipedalism from a dog's point of view. If they think about such things at all, dogs are probably far more impressed by our ability to stand up and throw sticks than by any of the achievements we might pick for ourselves, whether art, science, music or literature. The quintessentially human combination of manual dexterity and balance is unmatched anywhere else in nature. No other animal can juggle while standing one-legged on a tightrope! The easier tool and weapon use that results from this rather bizarre skill-set undoubtedly aided the survival of our ancestors, and modern humans owe much of their success to their ability to stand on two legs to use bows, bolas and boomerangs. Much of our culture, too, is centred on our ability to effortlessly free our hands for pot-making, drumming or computer programming. But, if you wanted to design from scratch an organism able to use her hands and brain to transform the world, wouldn't you give her at least four legs and dozens of arms? Why are we bipedal and bi-armed? The answer, of course, is that we have evolved from a quadruped. It might have been better to have had six-legged insects or eight-legged spiders for ancestors but those options too would have brought their own historical baggage in the shape of an inability for such animals to grow large enough to own complex brains. Quite simply, our current body structures are constrained by what could reasonably be done by evolution acting on the body structures of our ancestors. The resulting compromises are a clear signature that we are the product of an evolutionary process.

The message of this chapter thus far is that natural selection's ability to adapt organisms is impressive but not unlimited. The limits may be different on other worlds if their biochemistry, evolutionary history and environment are very unlike the Earth's but, even so, there will be limits. My examples have necessarily been restricted to Earth-based life, but, as the principle of mediocrity requires, our starting assumption should in any case be that our planet is reasonably typical of inhabited worlds (if this is not true, then my case is already made – the Earth really is special!). It is therefore reasonable to suggest that complex biospheres are not possible on all worlds. The environment in many places will simply be too difficult for life to begin, or if it does begin, too difficult for life to evolve large, complex organisms. The question, of course, is whether planets as life-friendly as the Earth are a very small set or a relatively wide one. However, when it comes to planets capable of supporting intelligent observers, I'm pretty sure that only a small minority of worlds are suitable. Partly this has to do with climate stability, which is the topic of much of this book, but there is in addition an intriguing puzzle whose most likely solution implies that the evolution of intelligence is difficult and rare. Why has intelligent life taken a similar amount of time to emerge on Earth as there was time available – 4 billion out of the 5 billion years over which the Earth has been and will be potentially habitable? As the next few paragraphs explain, the simplest answer is that intelligence is a very difficult trick to pull off.

Life began around 4 billion years ago and is unlikely to survive beyond another billion because of the inexorable warming of our Sun, which, at some point between 500 million and 1 billion years from now, will overwhelm our climate system. Our planet will then rapidly transform into something very like Venus and become uninhabitable. So, how do we explain the remarkable coincidence that the timescale for the emergence of intelligence is almost the same as the timescale for habitability? It could be that the evolution of intelligence really does take about as long as the time it takes for a star to start overheating, but this would be odd because they are very different processes. It would be a bit like noticing that rainfall in London depends on sushi consumption in Tokyo; there's no obvious link so why should they go together? Solar evolution is governed by nuclear processes and life's evolution is governed by biological processes; these are about as closely linked as the English weather is to Japanese eating habits. Thus, the similarity between the time taken for intelligence to emerge and the time permitted by our ever-warming Sun is pretty surprising. This problem has been analysed in detail by Brandon Carter, the Australian cosmologist who invented the term 'anthropic', and more recently by Andrew Watson from the University of East Anglia in the UK. Their explanation for the coincidence is that the true timescale for the emergence of intelligence is far longer than the timescale for stellar evolution.

An analogy may help here: professional football matches where ten goals are scored are extremely rare because, at an average goal-scoring rate, it takes much more than 90 minutes of football for that many goals to be scored. Furthermore, on the rare occasions when it does happen, the tenth goal is far more likely to be scored towards the end of the match than near the beginning because it is even less likely that you'll get all ten goals in, say, 45 minutes than that you will get them in 90 minutes. If you don't believe this, take a look at some recent football results. In the last five seasons of England's top two leagues there have been only four ten-goal games and in three of them the last goal was scored within ten minutes of the end (in the other match it was eleven minutes from the end).

Like high-scoring soccer matches, intelligent life hardly ever happens because there simply isn't enough time for all of the intermediate goals (origin of life, origin of photosynthesis, origin of complex cells, origin of animals, origin of intelligence, etc.). But, on the very rare worlds where intelligence does by chance evolve quickly enough, it is likely to appear shortly before the planet becomes uninhabitable. The analogy also demonstrates why life is likely to emerge early on these planets, as I discussed in Chapter 1. If that first goal doesn't come along for a long time, that reduces the remaining time available for the following nine goals. In three out of the four high-scoring matches discussed above, the first goal was scored within ten minutes of the start of the match.

The emergence of intelligent life on Earth is like a ten-goal football match: the first goal (origin of life) was scored early to give the maximum possible time for the remaining steps and the last goal (origin of intelligence) came towards the end. Note that, just because a goal is scored near the start of a match, that doesn't mean the match will have ten goals; it just improves the chances of it happening. Similarly, an early start for life doesn't guarantee the eventual emergence of intelligence but it does improve the likelihood (from minuscule up to tiny).

Before I move on, there is one further point worth making about this idea that intelligence will nearly always emerge 'at the last minute' if it emerges at all. The detailed analyses undertaken by Carter and Watson showed that each of the unlikely steps needed to take life from its origin through to the emergence of intelligence should be approximately evenly spaced through the available time. In a similar way, if ten goals are scored in a football match they will tend to be roughly evenly spaced through the game rather than having, say, one at the start and nine in the last five minutes. Even spacing of life's tricky steps has two important consequences. Firstly, the length of time taken for the first difficult step will be very roughly the same as the gap between the last difficult step and the end of the world. And that is exactly what we see on Earth, with the origin of life happening within 500 million years of suitable conditions appearing and ourselves appearing when life has about 500 million years left to run. Secondly, given 5 billion years of habitability and 500 million years between difficult steps, there must be roughly nine difficult steps. I should add that Carter's and Watson's estimates for the number of steps were lower than nine because we now think that the origin of life was earlier and that the end of the world will be sooner than either of them assumed. It should also be made clear that none of this implies that 'intelligence' is the goal of 'evolutionary progress'. We'd get the same answer if we looked at, say, the evolution of fruit. This too requires unlikely steps that will not happen on the vast majority of suitable planets. It's just that oranges don't think about this sort of thing.

So, it would seem that it typically takes a long time for life to become established and for it to evolve to the point where organisms as complex as human beings or citrus trees are possible. Under these circumstances, one of Earth's few indisputable oddities becomes very surprising indeed: the large size of our Sun. As I mentioned in Chapter 1, 95 per cent of stars are less massive than our Sun and, because the smaller red dwarfs burn their nuclear fuel slowly, they are long-lived and slowly evolving. Planets orbiting such stars at the right distance will therefore stay at a habitable temperature for much longer than the Earth: trillions of years in the case of the smallest stars! Intelligent life on these worlds has much more time to evolve and so should be much more likely to appear. If that's true, why do we inhabit such an unpromising location? Perhaps our planet is an oddity even among inhabited worlds, but it is more likely that there is something unpleasant about living near a red dwarf. Small stars are, for example, more prone to producing large stellar flares and, to be warm enough, habitable worlds will need to be close to such stars and their flares. Perhaps the size of our Sun is a compromise between the dangerous flares of smaller stars and the very short life of larger ones.

In this chapter I've tried to show that, though evolution's achievements are extraordinarily impressive, there are nevertheless significant limitations to what it can do. Life cannot adapt to absolutely anything that is thrown at it. In particular, large temperature changes that fall outside the range an organism would normally expect to encounter in its lifetime are especially dangerous. Indeed, as we've see in earlier chapters, occasional big changes in the Earth's climate have led to mass extinction events when the majority of living species died out over a geologically short space of time. Fortunately, things never got so bad that life was completely wiped out. And though it remains true that natural selection has played a central role in the success of life on Earth, environmental conditions on our planet, and temperatures in particular, might not have remained within the limits with which evolution can cope. The surprisingly stable climate over Earth's history still needs explaining. It's time now to take a proper look at the Gaia hypothesis to see if that can resolve the mystery of 4 billion years of good weather.

10
Pond Weeds and Daisies

Aquatic weeds are a nuisance. They're irritating enough when they wrap themselves around my rudder when I'm dinghy racing, but more seriously, they also cause substantial ecological damage when they take over too much of a river system or lake. Even this, though, is a very minor problem compared to events 49 million years ago when a single species of pond weed may have brought an end to the warm climate our planet had been enjoying and kick-started global temperatures on a downward spiral towards the cold and highly glaciated planet we inhabit today. Prior to this catastrophe the early Eocene Epoch, which began about 56 million years ago, was a time of high carbon dioxide levels and very warm temperatures. Indeed, this period of time is possibly the best ancient parallel for the world humans will create if we persist with our greenhouse gas emissions. However, after about 7 million years of warm weather, carbon dioxide levels and temperatures started to fall. Many explanations have been advanced for why this cooling began but one of the most interesting involves weed infestation on a truly monumental scale.

All the action took place in the Arctic. During the Eocene the Arctic was a very different place from the frigid wasteland of today because the high carbon dioxide levels allowed the polar region to be about as warm as modern France and at least as wet as the lush and famously emerald-tinted island of Ireland. The heavily forested lands that then fringed the ancient Arctic Ocean were so warm, even in the permanent dark of the polar winter, that there were lizards, tortoises and alligators living inside the Arctic Circle. The geography too was rather different from today, since the Arctic Ocean's link to the North Atlantic was then much narrower and, on the other side of the ocean, Alaska and Siberia were joined together to close off the Bering Strait. The Arctic waters were almost completely isolated from the rest of the world's oceans and this allowed fresh river water and rain water to accumulate on the ocean surface as a light, brackish layer floating on denser, saltier water a few metres below. Even today the upper 50 metres or so of the Arctic is less salty than water in the rest of the world's oceans, but in the Eocene, the Arctic was more isolated than it is now, rainfall was heavier, and stirring by waves was less intense so that this effect was greatly magnified. As a result, the Eocene surface layer on occasions became completely fresh. Those were the conditions that allowed rapid growth across the Arctic of *Azolla*, a freshwater fern normally incapable of living on a salty sea but able to thrive on this very odd, almost salt-free ocean.

Azolla is a fascinating plant. Unlike the land-loving ferns most of us are familiar with, *Azolla* grows as free-floating mats on the surfaces of lakes and rivers. It can do this because of its symbiotic relationship with bacteria that are able to break apart the strong molecular bonds in atmospheric nitrogen to make more plant-friendly compounds. *Azolla* carries these useful companions around in specially formed nodules that therefore provide it with the nitrate fertiliser that a free-floating plant can't get by the usual approach of sticking roots into soil. In return for this invaluable service, the plant provides its bacterial guests with food in the form of glucose and this partnership allows *Azolla* to thrive in nutrient-poor waters where other aquatic plants struggle. In the case of the Eocene Arctic Ocean with its very unusual freshwater surface layer, this ability occasionally allowed ocean-sized *Azolla* blooms to form in the polar summer. The northern polar seas were then covered by a mat of vegetation from the coast of North America to the shores of

Russia on the far side of the ocean more than 2,000 kilometres away. The entire Arctic Ocean became plant-green rather than the ice-white you might have expected. But in the long, dark polar winter, these weeds died or possibly just became dormant and much of the vigorous summer growth sank into the deep ocean. This could well have been the cause of the mid-Eocene cooling, because, as discussed in an earlier chapter, large-scale burial of plant matter in ocean sediments removes carbon dioxide from the atmosphere. Levels of the greenhouse gas would have fallen as *Azolla* remains accumulated at the sea floor and the resulting temperature drop may have been large enough to begin the descent towards the icy world of the last 2.5 million years: a world of regular ice ages interspersed by relatively short but warmer interglacial phases such as the one we have been living through since about 9000 BC. Many other factors played important roles in these changes to global climate but *Azolla* blooms may have been the straw that broke the camel's back – a final trigger that set off a complex chain of events flipping the world from the warm climate of the early Eocene into the colder climate of more recent times.

This '*Azolla* event' was, geologically speaking, a brief episode lasting little more than a million years. Even during that period *Azolla* blooms occurred rarely and seem to have been associated with occasional drops in global sea level that cut the Arctic Ocean off from the rest of the world even more than it already was. The world's oil industry is currently very interested in exploiting the hydrocarbons that can be found in the Arctic Ocean and an *Azolla* layer, which is one of the principal sources for this oil, has been found everywhere that drills have penetrated marine sediments of the right age. This discovery of the remains of a freshwater plant across an entire salt-water ocean was more than a little surprising, but the evidence for ocean-scale blooms of this particular aquatic weed is strong. The sediments show too much *Azolla* detritus spread across too wide an area for it to simply be the result of run-off from the land. The sediments also show no other terrestrial remains, such as wood fragments, that would be expected in land-derived deposits.

The *Azolla* event was a rather unusual chapter in Earth's history, but it is far from being the only time when there may have been life-generated climate change. Perhaps the best known additional examples are the Proterozoic cooling associated with the oxygenation of our atmosphere, discussed earlier, and the cooling that occurred 400 million years ago when roots of the newly evolved land plants accelerated weathering of rocks and increased the rate at which carbon dioxide was removed from our atmosphere. Events like these leave little doubt that living organisms have had significant effects on our climate but are these effects beneficial, catastrophic or some mixture of the two? Whatever the answer to this question, the impacts of life on climate are important to this book, because they almost certainly play a key role in the central problem I am tackling: What has kept our climate suitable for life throughout the 4 billion years of life's existence?

In previous chapters I've tried to show that, even for worlds that begin well, there are no guarantees that climate will remain life-friendly for billions of years. As Mars and Venus both show, planets with initially fairly benign environments can later become too hostile to sustain a complex biosphere, even allowing for life's extraordinary adaptability. In contrast, the Earth has somehow remained habitable across the eons and, if anything, has become more life-friendly as it has aged. So, why has terrestrial climate history been so different to that of our planetary neighbours? There are plenty of small differences, as discussed in earlier chapters, that may account for this happy circumstance. The absence of magnetic fields, the slightly different distances from the Sun, and the smaller sizes of our sister worlds all seem to have a role to play. However, there is one additional and particularly striking difference between the Earth and its neighbours – life itself! Perhaps life helps to stabilise climate. If this is true, and if life never became firmly established on Venus or Mars, this would explain the better climate outcomes on our own planet. The idea that life helps to create and sustain conditions suitable for life is, as I've mentioned before, called the Gaia hypothesis and much of the rest of this chapter will be concerned with this fascinating idea.

When it comes to discussing Gaia, everyone starts with a world full of daisies because there is no better illustration of how Gaia might work than 'Daisyworld'. This imaginary planet was invented by Gaia's originator, James Lovelock, and his collaborator Andrew Watson who we met in the last chapter. Daisyworld is a wonderfully simple hypothetical planet inhabited solely by daisies of just two varieties: black and white. Black daisies are better than white daisies at absorbing sunlight and so thrive when Daisyworld is cool, while the white variety stays cool in sunlight and hence thrives if Daisyworld becomes hot. As a direct result of this arrangement, Daisyworld becomes darker in times of cold climate, as more dark flowers grow across its surface, and lighter when it's warm as the overheating black daisies die off to be replaced by their paler cousins. The resulting global colour changes stabilise Daisyworld's climate, since the darker planet of cooler times absorbs more heat from the Sun and a whiter world reflects more heat into space when the climate becomes warmer.

While Daisyworld is an inspiring idea, no one pretends that it's anything but a greatly simplified toy-model designed to illustrate a concept rather than realistically represent the behaviour of the highly complex real Earth with its millions of interacting species. However, it's quite easy to come up with equally simple alternative hypothetical planets that behave in the opposite manner. Take 'Mouldy Pineworld' for example. Like Daisyworld, this planet has just two species: pine trees that turn carbon dioxide into wood, and mould that grows on the trees. Mould breathes oxygen in and carbon dioxide out and grows better in the warm, so on Mouldy Pineworld an increase in temperature encourages mould growth, which thrives at the expense of its hosts. The growth then increases the amount of carbon dioxide in the atmosphere and this, in turn, increases temperatures yet further. A cooling of Mouldy Pineworld, on the other hand, kills off the mould and encourages tree growth, which decreases carbon dioxide levels and enhances cooling. In contrast to Daisyworld, Mouldy Pineworld suffers from positive feedback, and so its biosphere magnifies rather than moderates changes in climate.

Daisyworld and Mouldy Pineworld are both imaginary, but one genuine biological climate feedback found on Earth, the response of northern forests to warming or cooling, is very nearly a combination of Daisyworld and Mouldy Pineworld – and it shows that both are wrong! The average temperature of our world has changed repeatedly over the last 2.5 million years, as the ice ages have cycled through the long glacial and short interglacial periods, and the coniferous forests of northern Europe, Asia and America have responded to this by spreading north during warm times and retreating southwards when the cold returned. This is the exact opposite of Mouldy Pineworld, where the trees spread when it was cold rather than when it was warm. On the other hand, as the forests moved northwards they grew over what was formerly tundra and those invading trees were substantially darker than the frozen wastelands they replaced. Thus, as the world warmed, the spreading northern forests made the Earth darker and more heat-absorbent. This is the exact opposite of Daisyworld, which became lighter as the world warmed. The climate response of the northern forests therefore beautifully shows how misleading simplified models can be if they are taken too seriously.

My invention of Mouldy Pineworld was, of course, highly contrived to ensure that it showed living organisms interacting in a way that necessarily produced the climate instability I was trying to demonstrate. However, Daisyworld is just as vulnerable to a criticism of being highly contrived, because the petal colours it assumes are far from inevitable. Black is indeed the warmest colour in bright sunshine, but it is the coolest colour at night or on a cloudy day. Dark colours are best both for gaining heat in sunshine and for losing heat in the shade, which is why, unlike mad dogs and Englishmen, many inhabitants of hot countries wear dark clothes and then stay out of the mid-day sun. I saw another example of this basic physical property of dark colours a few days ago as I was travelling by train through the snowy countryside of a wintry, but slowly thawing, Essex. As I looked out of the window I was struck that some fields remained blanketed in white while immediately adjacent fields had completely lost all snow. It took me a few minutes to work out what had produced this remarkable dichotomy but the key was that some fields had been ploughed and others had been left as grassy pasture. The relatively dark colour of tilled ground where it stuck out above the snow-filled furrows had allowed it to cool more overnight and the snow had not yet melted from these slightly colder fields. As these examples show, black is often the worst colour for staying warm and so it is far from obvious that dark daisies would have an overall survival advantage on a cold Daisyworld. If white daisies thrived instead in chilly weather, disaster would ensue. Daisyworld would become whiter as the climate cooled and this would enhance rather than retard cooling. The resulting anti-Daisyworld would look and behave very like the snowball Earth we met earlier in this book; a world that lurched uncontrollably between a brilliantly white but near-lethally cold state and a darker but near-lethally warm one.

Daisyworld can also be modified to illustrate another difficulty that the supporters of Gaia are working hard to resolve: the problem of cheating discussed in Chapter 9, which arises because all of life's adaptations have costs as well as benefits. The human brain is an excellent example of an adaptation with both good and bad effects, because its development was key to our success as a species but it requires significant extra food to grow and maintain such an extravagant organ. A little less dramatically, there is also a cost to producing black or white pigment if you are a daisy. The biological cost of pigment production comes because it takes nutrients and sunlight to produce them, resources that could have been used for other purposes such as growing. This cost causes a problem for Daisyworld because, under such conditions, it would be beneficial for daisies to cheat at the natural selection game in exactly the sense I discussed in Chapter 9. Specifically, a mutant, unpigmented daisy can keep itself warm in cold times by growing next to sunshine-absorbing black daisies or cool in warmer times by growing next to sunshine-reflecting white daisies because the pigmented flowers affect not only their own temperature but also that of the surrounding air. Unpigmented daisies therefore benefit from the climate control of their neighbours without paying the cost of producing pigmentation. These cheats will out-compete their pigmented cousins even when the black and white daisies become rare, since they will still benefit from any marginal remaining climate control. Eventually, pigmented daisies might even be driven to extinction and Daisyworld would lose its Gaian stability.

In fairness I should admit that there is an ongoing and mathematically sophisticated scientific debate about whether the climate-stabilising behaviour of Daisyworld really is destroyed by such cheating but, nevertheless, the discussion does illustrate a genuine problem for Gaia's supporters: a problem widely known as 'the tragedy of the commons'. This parable was originally devised to illustrate a fundamental conflict between laissez-faire economics and good husbandry of our planet but it equally well demonstrates why anti-Gaian tendencies might be expected to evolve in ecosystems. Common lands, in England, are shared fields where all the inhabitants of a village have grazing rights. In the tragedy of the commons it is imagined that a particular field can support ten cattle and that ten villagers are entitled to use it. It would therefore make sense for the farmers to graze only one cow each, but a selfish villager could gain a distinct economic advantage over his neighbours by grazing two cows. There would then be eleven marginally malnourished cattle on the common, but because two slightly hungry cows are worth more than a single well-fed one, the selfish farmer would do better than his more community-minded colleagues. The other farmers might attempt to compensate for the reduced value of their own cattle by also grazing two cows each and, at that point, the common would become so over-grazed that all the cows would starve. Every villager (let alone their cattle) would end up worse off than if they had agreed among themselves to stick to a single cow each. Entirely rational, if selfish, behaviour by an individual produces catastrophe for all.

Humans are, at least in principle, capable of realising what is happening and of making agreements to prevent such insane over-use of limited resources. A good example is fisheries policy in which countries agree to limit the amount of fish they each take from the ocean rather than having a free-for-all that drives fish stocks to extinction. However, the same kinds of problems also occur within evolving ecosystems and, in this case, the uncompromising rules of natural selection do not permit agreements that limit the behaviour of individuals in the interest of everyone. Natural selection takes place at the level of single organisms and there is no mechanism to allow selection of attributes 'for the good of the species' let alone for the good of an entire ecosystem. This deeply anti-Gaian principle at the heart of life's dynamics is well illustrated by a very stark real-world example: the near-absence of life across most of the Earth's surface.

We are used to thinking of the Earth's oceans as teeming

with marine organisms but the open ocean is actually quite desolate and typically has about twenty times less biological productivity than those parts, such as coral reefs, that tend to get shown on television. The problem is that the seas are over-grazed in exactly the manner discussed for the common land above. In the ocean's case the grazers are microscopic organisms rather than cows and they have over-consumed nutrients rather than grass. Iron, for example, is in very short supply in many ocean waters since it can only get a long way from the coast by being blown off the mineral-rich land in dust particles. Unfortunately, a little bit of dust doesn't provide much metal and there is increasing evidence that photosynthetic marine micro-organisms compete for what little iron there is – and they may even hoard it. Any easily accessible iron dissolved in the sea is therefore rapidly removed to leave an environment that cannot support the biodiversity it might otherwise have sustained. The contrast between the barren oceans covering much of the Earth's surface and those parts of the ocean fortunate enough to receive nutrient-rich water upwelling from the dark, and photosynthesiser-free, waters of the deep sea is striking. All the world's major fisheries are either in areas of upwelling or close to land where nutrients can be supplied by air-blown dust.

The effect of iron deficiency on ocean life also plays a role in another interesting example of the multi-faceted interactions between life and the physical environment: the role of marine aerosols in our climate. Aerosols are solid particles or liquid droplets in air that are small enough to remain suspended for long periods. They undoubtedly play a direct role in climate by making the air less transparent, thus reflecting solar heat back into space and Earth heat back to the ground. Furthermore, aerosols play an indirect role by 'seeding' clouds. They provide particles around which water vapour can condense more easily than it does in their absence. Aerosols are produced by many processes such as volcanic eruptions or human activities, or simply by spray from the sea. The suggestion that sea-spray might, indirectly, encourage rain was first made by Glen Shaw, a geophysicist at the University of Alaska who had noticed that even the pristine atmosphere of the Antarctic has a very slight sulphurous haze that he later tracked down as originating from chemicals sprayed by breaking waves into the atmosphere. Shaw suggested that this could produce a significant negative feedback in the climate and his ideas were expanded on in a classic scientific paper outlining what is now known as the CLAW hypothesis (named after the paper's authors Robert Charlson, James Lovelock, Meinrat Andreae and Stephen Warren). In this proposal, the marine aerosols contain a chemical called dimethyl sulphide (DMS), which is generated by microbes in the ocean. The DMS aerosol helps clouds to form and might be produced in larger quantities on a mild world with less ice cover and warmer, more microbe-friendly water. Thus, clouds should form more easily when the Earth is warm and these could act to reflect heat into space, counteracting the warming.

This was a nice idea, but further work has thrown up three major difficulties for the hypothesis. Firstly, the microbes are held back not by cold weather but by the iron shortage I discussed earlier. DMS is produced in greater quantities when it is dusty, and detailed examination of ice cores from Antarctica and the Arctic show the Earth to be dustier during ice ages than it is during warmer interglacials. More microbially generated DMS is therefore produced when the world is cold than when it is warm. The second problem is that clouds have a complex effect on the climate and frequently warm it rather than cool it. Clouds can hold warmth in as well as reflect sunlight away, and that's why, for example, cloudy winter nights are warmer than cloudless ones. It is also now thought that much of the DMS is associated with sea salt rather than microbes, so it's not even clear that biological productivity is the main control on its atmospheric concentration. So, once again, we see that the relationship between life and climate is complex and difficult to untangle. The DMS story may illustrate a negative feedback in the climate but it could equally well be a positive feedback. It may even be a negative feedback for exactly the opposite reason to that given by CLAW. DMS-generated clouds may be more common in cold times than warm times but may have a warming rather than cooling effect!

To me, the picture that emerges from all this is that the Gaia hypothesis lacks unambiguous observational support and has significant theoretical difficulties. That seems pretty damning but I'd nevertheless regard my views on Gaia as being broadly positive. My attitude can, I hope, be gleaned from events surrounding a poster entitled 'Gaia or Goldilocks?' which I presented at a conference concerning 'Life and the Planet' in London a few years ago. The text for my poster began:

Pseudo-Gaia

Imagine that everything you think you know about Gaia results from coincidence. Every cancellation of a physical effect (e.g. an inexorably warming Sun) by biological evolution (e.g. the rise in atmospheric O_2) happened by luck at just the right time rather than because Gaia required it. Phenomenally good fortune would be needed but there are $\sim 10^{22}$ planets in the observable Universe (and vastly more beyond our cosmic horizon) so that even highly improbable things will happen sometimes. Imagine further that a stable environment is a necessary precondition for a complex biosphere and that a complex biosphere, in turn, is necessary for the emergence of observers. Observers would then only emerge on rare worlds where all the coincidences required for a stable environment have occurred. How would the World look to such observers? It would look like Gaia and would have always behaved like Gaia, but it would not be Gaia. Is this our World?

As I stood by my poster discussing its contents with other conference attendees, I was approached by a sweet little old lady who politely enquired whether my poster might be 'a little discourteous'. 'Oh, I do hope not', I replied, 'it's certainly not meant to be.' Despite this reassurance I could see that, on second thoughts, producing a poster with the sub-title of 'Pseudo-Gaia' at a conference organised largely by Gaia enthusiasts might indeed be seen as unnecessarily aggressive so imagine my horror, half an hour later, when that sweet little old lady climbed onto the lecture theatre stage to give a conference keynote talk. My interlocutor had been Lynn Margulis, at the time one of the world's greatest living biologists (sadly, she died later that same year) and a long-term collaborator with James Lovelock on Gaia theory. My point here (and on my poster) is that I regard the Gaia hypothesis as a fascinating and productive idea despite my opinion that it is probably wrong. Indeed, at the end of this chapter and in the next one, I will discuss a mechanism that I believe really could produce Gaian biospheres on inhabited worlds. However, before that, I want to look at a rather different, non-Gaian, way in which life could play an important role in keeping the Earth life-friendly.

I think that what is needed is a slightly different perspective on the problem of how biology, geology and astronomy interact to produce a broadly stable climate. Perhaps the emphasis on feedback, whether positive or negative, is misplaced. Perhaps an evolving planet naturally imposes a general trajectory on climate: a tendency to cool a planet and hence counter the warming influence of solar evolution. Steady continental growth, for example, has slowly increased the area available for weathering and also represents a permanent store for the carbon.

Furthermore, David Schwartzman, a professor from Howard University in Washington, DC, believes that life's major innovations have tended to reduce greenhouse gas warming. An important example for Schwartzman is the emergence of primitive lichens that colonised the continents and helped to break down the rocks that they grew upon, perhaps as early as 700 million years ago. This increased weathering would, as we have seen, have led to more rapid removal of carbon dioxide from the atmosphere as acid rain dissolved the fragmented rocks and flushed the resulting bicarbonate ions into the oceans. Colonisation by lichens also had the effect of opening up the continents for other life forms and this general trend for the biosphere to expand through time is another central feature of Schwartzman's proposals. Land plants, when they emerged hundreds of millions of years later, increased the rate of weathering and the degree of colonisation still further and so ratcheted up cooling by another notch. These examples illustrate a tendency for life to evolve a succession of increasingly efficient mechanisms for conquering the land, at each step removing more carbon from our atmosphere to give the otherwise quite unexpected downward trend in temperatures seen through geological history.

Micro-organisms, lichens, fungi and plants certainly strongly enhance weathering by helping to break up the solid rocks of the Earth and so it is very possible that the weathering feedback mechanism would have been too weak on a lifeless Earth to help stabilise the climate. It is also quite probable that more complex organisms could not have evolved unless increasingly efficient weathering, due to increasingly effective land organisms, had allowed global temperatures to fall. Continental growth and successively more efficient species of rock-digesters (e.g. trees) therefore acted to bring temperature down. We could be living on a world where the warming effect of an evolving star happened to be roughly cancelled by the cooling effect of geology and biology with no overall control from feedback mechanisms at all.

The best evidence available to support this idea of broad cancellation of solar heating by biological cooling comes from the well-constrained history, discussed earlier, of equatorial sea surface temperatures over the last half-billion years. As I mentioned, these sea surface temperatures show a very slight cooling of a couple of degrees centigrade superimposed upon more random, background fluctuations of about 10°C. Feedback, whether resulting from geochemical or biological processes, cannot explain the overall cooling, because feedback dampens change but cannot reverse it and so at best could only have reduced the expected warming of about 10°C that should have been produced by the steadily evolving Sun. In contrast, the combination of two opposing processes that roughly cancel each other would leave a small residual temperature fluctuation that could be either minor warming or minor cooling. On our world it just happens to be a slight cooling.

Of course, this rough cancellation requires a coincidence. The amount of cooling has to roughly equal the amount of warming and there's no obvious reason why this should happen. Why should the timescale and amplitude of biological cooling be anything like the timescale and amplitude for the very different and completely independent processes causing solar warming? Schwartzman's answer would be that this is Gaia in action but, it seems to me, there is a much simpler and more straightforward explanation. If bio-geological evolution occurs at different rates on different worlds, intelligent observers will emerge only on those planets that, by chance, have just the right rate of geological and biological cooling. This is the anthropic principle again, the observational bias that inevitably results from the fact that the history of our planet must be compatible with our existence as observers. On this interpretation, such cancellation does not occur on the vast majority of Earth-like planets. The usual situation is either that biology freezes itself to death or that solar physics fries the world well before complex organisms ever evolve. We just happen to be living on one of the rare, lucky worlds where neither of those things happened.

I should add that combining Gaia and Goldilocks in this way is not really a new idea. Rather, it is a twist on a proposal called 'Lucky Gaia', which suggests that only those planets that, by chance, produce Gaian mechanisms will give rise to observers. However, to me, this idea rips the heart out of Gaia. Is it really Gaia at all when the anthropic principle has done all the heavy lifting?

This is all, perhaps, a little too negative and I can hear the spirit of Professor Margulis politely requesting that I tone things down a little. So, I'd like to finish this chapter on a more positive note by discussing one possible way forward for Gaia. For all the reasons I have been through above, I have difficulty believing that Gaia can emerge simply as the result of a preponderance of negative feedback over positive feedback among the innumerable interactions that occur in our biosphere. Rather, I agree with those who have suggested that Gaian stability can be understood only as an emergent property of a very complex system. Emergent properties can be illustrated using a heap of sand as an analogy. If you slowly add sand to a pile, one grain at a time, its height gradually grows along with the slope of its sides. However, at a critical point, the sand pile stops getting any steeper and any additional grains simply slide down the sides to make the pile wider rather than taller. The maximum slope on the sides of this pile is a property of the sand but not a property of a single sand grain. It simply doesn't make sense to talk about the maximum stable slope for a single grain – and, indeed, this is a property that doesn't emerge until there are at least a few tens of grains in the heap. The maximum stable slope is therefore an emergent property of sand grains. Similarly, environmental stability may be a property of ecosystems that emerges only once there are large numbers of interacting biological and geochemical processes. On this interpretation, the life-generated stability of our environment simply cannot be understood as the crude sum of individual species–climate feedbacks.

However, why should the laws of nature do this? It seems just too good to be true. Personally, I can think of only one plausible answer and it comes back, once again, to the anthropic principle but now played out on an immensely larger stage. Perhaps we live in a lucky Universe, rather than on a lucky planet. Perhaps the entire cosmos is fine-tuned for life.

11

Life's Big Bang

Before the 1980s most scientists found the suggestion that we live in a Universe fine-tuned for life outrageous and implausible. The one and only Universe is what it is and it could be no other way. How could a Universe, in any sense, be optimised for life? This attitude has gradually changed over the last 30 years until the idea that we live in a peculiarly life-friendly Universe has now almost, but not quite, become mainstream cosmology. During this time, cosmologists have developed weird and wonderful ideas in an extremely successful attempt to explain the observed large-scale structure of our Universe. These same ideas, almost incidentally, make cosmic fine-tuning for life not just plausible but almost inevitable. This chapter therefore looks at this cosmological background and how it may impact on my central topic: the life-friendliness of the Earth.

A couple of years ago, after giving a talk on earthquakes at a local school, I was asked: 'What's the most surprising scientific discovery in your lifetime?' I had no doubt about the answer but had to confess that it was in cosmology rather than my own subject of geophysics. It's been known since the late 1920s that distant galaxies are moving away from us as a result of the Big Bang that created our Universe but, 70 years after the discovery that we live in an expanding cosmos, the scientific world was shocked by the revelation that the galaxies are spreading out ever more rapidly. It's like watching as a ball thrown into the air accelerates away into space instead of slowing and falling back to the ground.

The equations that describe evolution of the Universe have a lot in common with those that govern the trajectory of a ball.

Gravity decelerates a thrown ball as it rises until it eventually stops and then falls back, although in theory a ball could be thrown faster than Earth's escape velocity so that it would never quite come to a stop. Even then it would still slow as it travelled upwards – just not by enough to ever become stationary. In a similar way, gravity should slow the Universe's expansion. This deceleration could happen so rapidly that the Universe would eventually stop expanding and then contract. Alternatively, the expansion could slow more gradually so that it never quite ceased. Cosmologists in the second half of the 20th century worked hard to measure the rate at which the expansion is slowing, because they wanted to know the ultimate fate of the Universe. Would it turn around and collapse into a 'Big Crunch' or would it eventually disperse into nothingness? To everyone's surprise, they discovered a Universe whose expansion is not slowing at all. A mysterious repulsive force is pushing distant galaxies so that they move away from us ever more quickly rather than having their recession slowed by the force of gravity.

The approach used to determine the cosmic deceleration was straightforward in principle. Astronomers compared expansion in the relatively nearby parts of the Universe to expansion further away. Light from more distant parts of the Universe has taken longer to get to us and so, if we look at how fast distant objects are receding from us, we are seeing expansion at an earlier time. I should warn that 'nearby', in this context, bears no relation at all to what most of us understand by the word. To see the expansion at all we need to look at galaxies hundreds of times further away than our nearest significant extra-galactic neighbour, the famous spiral galaxy in Andromeda. Light has taken over 2 million years to get to us from the Andromeda spiral and hundreds of millions of years to get to us from the galaxies I'm describing as 'nearby'. Those galaxies are close, though, compared to the ones used to determine the ancient expansion rate.

Whether near or far, the speeds at which galaxies are moving away from us as the Universe expands are measured using the Doppler effect, the fact that the colour of light emitted by receding galaxies is shifted towards red as successive light waves come to us from further and further away. However, to measure expansion, we need to see how recession speed increases with distance, and we must therefore find out how far away each galaxy is. Measuring that distance is much more difficult than determining speed.

The most successful method for estimating the huge distances involved is to accurately measure the brightness of supernova explosions in distant galaxies. The further away these are, the fainter they look and so measuring their brightness tells us their distance. This is much harder than it sounds. There are several different types of supernova and only one kind has reasonably consistent luminosity. Furthermore, even these gigantic stellar cataclysms are hard to spot when looking half way, or more, across the visible Universe. A final difficulty is that these measurements must be calibrated using a supernova from a nearby galaxy whose distance is determined by some other technique. Cosmic distance scales are built up using a step-by-step approach in which we first determine the size of the solar system, then the distances to nearby stars, then distances to nearby galaxies and so on. Each of these steps is subject to uncertainties and so inter-galactic distances are hard to determine with the necessary precision. As a consequence of all these problems, the results from the supernova surveys were confusing and contradictory for several decades; but, when unambiguous measurements finally arrived partly thanks to the Hubble Space Telescope, the results astonished nearly everyone. The Universe's expansion had been slowing as expected during the first 9 billion years of its existence but, about 4 billion years ago, the expansion started to speed up again.

Although I've emphasised that this discovery was

unexpected, the theoretical possibility of a Universe with accelerating expansion had been known for more than 80 years. Few people took the possibility seriously because it requires anti-gravity, a repulsive force large enough to overwhelm the usual gravitational attraction between massive objects. Anti-gravity is a mainstay of much science fiction, and sounds like pure fantasy, but Einstein's very successful theory of general relativity actually allows it. In this theory, which was published in 1915, gravitational attraction is caused not only by mass (as in Newton's earlier theory) but also by pressure. The force of gravity pulling you towards the Earth right now, for example, results almost entirely from the large mass of the Earth, but in general relativity this force is minutely supplemented by an additional attraction caused by the immense pressures inside the Earth. In normal matter the contribution from pressure is utterly swamped by the contribution from mass and can be ignored. Nevertheless, the contribution is there. Furthermore, pressure, and therefore gravity, can be negative: for example, consider stretched rubber, which pulls inwards rather than pushing outwards. So, in principle, we generate anti-gravity every time we stretch an elastic band. In practice any real material would snap long before measurable effects occurred, but it turns out that space itself exerts negative pressure.

It may seem surprising that a vacuum exerts any kind of pressure, but this results from another major 20th-century breakthrough in physics: the laws of quantum mechanics. In quantum mechanics the energy contained by an object, such as the wound-up spring of an old-fashioned clock, does not have a precise value but instead fluctuates on short timescales because of the uncertainties that are an inherent feature of quantum theory. This uncertainty in energy applies to everything, even empty space. As a consequence I shouldn't really use the phrase 'empty space' at all, because (thanks to Einstein's equation $E=mc^2$ which says that matter and energy are really the same thing) these energy fluctuations cause space to be filled by a dynamic froth of short-lived elementary particles. These 'virtual particles' give the vacuum real physical properties including a minute density and a minute pressure. Moreover, since a larger piece of vacuum contains more virtual energy than a smaller chunk, you have to put energy in if you want to stretch space. This implies that stretching space is hard work (as well as hard to imagine) and so the vacuum resists being stretched. Hence, like rubber, empty space exerts a negative pressure. The anti-gravity from this turns out to be three times greater than the gravitational attraction generated by the vacuum's tiny density and so, overall, empty space produces a minuscule amount of anti-gravity. In a nutshell, the vacuum has a repulsive nature. It's as if space doesn't like itself and every part of it is gently pushing away every other part. The effect is tiny and utterly unable to produce measurable consequences on human scales of space and time but you can see the resulting repulsion by looking at the cumulative effect as it adds up across 10 billion light years.

Before moving on to the cosmic consequences of this negative pressure, it's probably worth saying a few words about my terminology here. In cosmology books you will not often see the phrase 'anti-gravity'. Rather, you will see more scientifically acceptable synonyms such as 'dark energy', 'quintessence' or 'the cosmological constant'. The last of these phrases has the longest history. The cosmological constant was originally introduced by Einstein himself because, when he developed his theory of general relativity, the expansion of the Universe had not yet been discovered and it was believed that the Universe must be static. Einstein introduced the cosmological constant into his equations as a way to produce a balanced Universe where a general repulsion spread through the whole of space is exactly cancelled by the gravitational attraction between massive objects. However, by 1930, it became clear that the Universe was expanding and so the fiddle-factor Einstein had included to allow a static cosmos was no longer needed. Einstein famously dismissed its invention as his greatest blunder but the cosmological constant never completely went away. Theoreticians occasionally resurrected it to try to explain various oddities such as a Universe that seemed younger than some of the stars it contained! However, the cosmological constant was rather frowned upon as not quite respectable until it came back with a vengeance following the late 20th-century discovery that cosmic expansion is speeding up. This observational confirmation of anti-gravity was almost immediately interpreted as being due to the repulsive vacuum that I described above.

Dark energy, on the other hand, is a more recent term and is more inclusive in the sense that it covers alternative explanations for accelerating cosmic expansion that are not strictly equivalent to a cosmological constant (quintessence is one of these). These alternatives have a similar effect, in that they produce a long-distance repulsive force, but they differ in their mathematical details. These differences are unimportant here, and for simplicity I will stick with the most widely discussed explanation for an accelerating Universe – that it results from a vacuum-generated cosmological constant.

The repulsive force produced by all this weird and wonderful physics is so small as to be almost immeasurable but it is competing with gravitational attraction generated by an average density for matter that is also tiny. Vast empty spaces between galaxies take up most of the Universe's volume, and even the galaxies themselves consist mostly of near-vacuums in the voids between the stars. There is a complication here, however. Galaxies are filled and surrounded by 'dark matter' (not to be confused with 'dark energy'), a currently unidentified fluid whose only physical effects are through the gravity it produces. Dark matter does not, for example, interact with light and that's why we can't see it. Nevertheless, it must be there since the way stars move within galaxies (and the way galaxies themselves move) can be explained only if there is far more matter present than we can see. In fact, there is six times more dark matter in the Universe than there is of the ordinary kind from which stars, planets and people are made. Despite this extra contribution to the amount of mass in galaxies and galaxy clusters, the average matter density of the Universe remains very small – equivalent to about one thousandth of a gram in a volume the size of the Earth. Furthermore, as the Universe has expanded, this low average density of matter has fallen further, and, with it, the tiny gravitational attraction holding our Universe together has fallen as well. The cosmological constant, on the other hand, really is a constant. It has not dropped as the Universe has expanded since it is a property of empty space itself. Thus, as the Universe has grown, the relative importance of the cosmological constant has grown with it.

This growth in anti-gravity's importance can be seen clearly in the supernova data, which shows us that, about 4 billion years ago, the Universe expanded to the point where repulsion by the cosmological constant finally cancelled out the diminishing attraction from matter. Since that moment, the expansion has been accelerating rather than decelerating. The observation that gravity and anti-gravity were in balance 4 billion years ago also allows us to work out the size of the anti-gravity effect. The cosmological constant produces a pressure that is one hundredth of one trillionth of the atmospheric pressure at the Earth's surface. When spread across the entire cosmos, though, the gravitational effects of this almost inconceivably small pressure are sufficient to gently tear the Universe apart.

All of which brings me to one of several deep cosmological mysteries at the core of this chapter: physics cannot explain why this vacuum pressure is so very small. A back-of-the-envelope calculation based on the expected quantum fluctuations shows that there should be a staggeringly large vacuum pressure roughly equal to the so-called 'Planck pressure' of 'one followed by 108 zeros' atmospheres. The difference between this unimaginably large theoretical pressure and the unimaginably small true value has justifiably been called the worst prediction in the history of science. The surprise is not that our Universe has anti-gravity, but rather that this repulsive force is so small. This problem was first identified in the 1960s and astronomers' response at the time was to assume that the cosmological constant must be exactly zero. This is an entirely reasonable reaction. If you can't see an elephant in your room it's probably because there is no elephant there rather than because a ridiculously small elephant is hiding under a speck of dust. Then, 30 years later, came the strong evidence of an accelerating Universe and, hence, a non-zero cosmological constant. There was, after all, a microscopic pachyderm in the room and it was going to take some explaining.

At this point, the anthropic principle and the life-friendliness of our Universe finally come back into my story. Assume for now (I'll justify this more later) that there are many universes and that the vacuum-fluctuation pressure is able to take different values in each one. Under these circumstances, the typical pressure in a typical universe might still be close to the Planck pressure but a few very rare universes might, by chance, have much smaller vacuum pressures. In high-pressure universes, anti-gravity will be very strong and they will blow themselves to pieces long before stars, galaxies and observers can evolve. Such universes will therefore be uninhabited. Life can appear only in highly unusual universes where the vacuum pressures are low enough to allow time for galaxies and their possible inhabitants to appear. In this view, our Universe is one of those infrequent but life-friendly universes.

This is the kind of problem the anthropic principle was originally devised to tackle: explanations for life-promoting coincidences that seem to be built into the entire fabric of the Universe. There are quite a few such coincidences, since the cosmological constant is not alone in having a value that seems fine-tuned to allow life. Many physical properties of our Universe would, if changed slightly, render it uninhabitable. The best known example of this is that the strengths of electrical and nuclear forces are just right to allow carbon to be among the most common elements. Carbon chemistry is basic to all life on Earth because carbon has a unique ability to form a large number of complex compounds. It seems likely therefore that carbon, along with water, will be essential to any chemically-based life and so a cosmos in which carbon was rare would probably be a cosmos without life.

Fortunately, a number of factors are fine-tuned to allow carbon to form easily. Carbon is created inside stars when they start to get old. In younger stars such as our Sun, the energy is produced by the nuclear fusion of hydrogen to make the next heaviest element, helium. However, when the hydrogen starts to run out, stars collapse and their central temperatures and pressure go up dramatically. This creates the conditions under which helium nuclei themselves can fuse and so the stars get a new, relatively short-lived, source of energy. At first two helium nuclei fuse together to form beryllium and this reaction occurs quite easily. However, beryllium is actually quite rare because it is very easy to add another helium nucleus to beryllium, creating carbon. Most beryllium that forms is rapidly turned into carbon. This reaction could, in principle, go a stage further by the addition of yet another helium nucleus to produce oxygen. However, this second reaction is much harder and so only a little carbon turns into oxygen. Thus, the main product of nuclear fusion in stars, once they have used up their hydrogen, is carbon. Eventually, some of this carbon is blown into the rest of the Universe by the strong stellar winds of ageing stars or, more dramatically, exploded into space by novae and supernovae eruptions. This is the stardust from which all of us are made.

If the relative strengths of the electrical and nuclear forces were slightly different, this story would change dramatically, because it would alter the ease with which different elements can be synthesised inside stars. A cosmos in which beryllium was the stable end-product of stellar fusion reactions would contain a lot of violent and short-lived stars and very little carbon. A universe in which carbon reacted easily to produce oxygen would be rather damp, since it would be full of oxygen and hydrogen, the ingredients of water; and in this damp universe carbon would be as rare as beryllium is in our own Universe. Interestingly, this story shows how the anthropic principle can be used to make predictions. Fifty years ago little was known about the details of these nuclear reactions. However, Fred Hoyle pointed out that carbon was 'an essential component of astronomers' and so carbon must be a major product of stellar nucleosynthesis. Furthermore, he argued that for carbon to be so common it must be much easier to convert beryllium into carbon than it is to burn carbon to form oxygen. All of these predictions were experimentally verified within a few years.

There are many other ways in which the laws of physics seem to be fine-tuned to allow our cosmos to be particularly life-friendly. Planetary orbits, for example, are stable only in three dimensions but the latest theories in physics suggest that the cosmos actually has eleven dimensions, most of them too small to see. In our Universe, only three spatial dimensions took part in the Big Bang, but other universes may have different numbers of 'large' spatial dimensions. In another example of our life-friendly cosmos, molecular bonds have the right strength to allow chemistry at temperatures corresponding to typical star–planet separations. If the electromagnetic forces in our cosmos were much stronger, then planets would be too cold for chemistry; if they were weaker, the entire Universe would be too hot.

None of these properties have values set, as far as we know, by fundamental laws of nature, and quite moderate alterations produce imagined universes with substantially less complexity than our own – universes where galaxies, stars, planets, molecules, atoms or even nuclei cannot exist. Life is, if nothing else, complexity-unmatched and it stretches credulity to suggest that universes that can't even make atoms could somehow generate life. It's conceivable that the values of physical constants in our Universe are the only ones possible, but the alternative explanation, that multiple universes are actually realised in nature and that we necessarily live in one of the few where the constants are 'just right', is surely a fascinating idea worth pursuing.

There is much more that could be said about cosmological anthropic effects than has been discussed in detail by cosmologists themselves. The life-friendliness of our planet is not peculiar at all if, instead, it's a consequence of living in a life-friendly universe. For example, the combination of silicate rocks, carbon dioxide, and water stabilises the climate of Earth-like planets, as discussed earlier in this book, and this would not happen if the properties of any of these compounds were altered. Is this an example of a peculiarly life-friendly property of our particular universe? Could there also be fine-tuning to optimise the occurrence of habitable planets in our Universe or to make the long-term stability of climates more likely? Might our 4 billion years of good weather therefore simply result because we live in a universe where that sort of thing happens?

Can we go even further? Are the laws of nature in our Universe contrived not just to allow complexity but to actually make the emergence of life inevitable on suitable worlds? Could the Gaia hypothesis even be true, despite my scepticism in the last chapter, because a life-friendly universe is a universe that generates Gaian biospheres? The final two chapters of this book will tackle the issue of whether we live in a life-friendly universe or on a life-friendly planet; for now, let me pursue the cosmological version of the anthropic principle a little further.

The idea that we live in an unusually life-friendly universe makes sense only if two things are true. Firstly, there must be many universes. Secondly, the laws of physics must be different in each of them. Given this situation, some universes will be better suited to life than others, and it is then almost inevitable that we will find ourselves living in one of the better neighbourhoods. Note, however, that for the same reason I suggested earlier that we probably live on the second-best of all possible worlds, we probably also live in a second-best universe. Universes that are marginally suitable for life will probably be far more common than universes that are perfect. Nevertheless, multiple universes each with different laws of nature explains a great deal about our Universe that otherwise seems mysterious. But is the idea of a collection of universes, a multiverse, really believable?

The simple answer is 'yes' because, as many teenagers might argue, our part of the cosmos is simply too boring to be all that there is to existence. Let me explain this rather cryptic statement. It has been understood for many decades that the very smooth and featureless character of our Universe, on the very biggest scales, is difficult to explain. On cosmologically small scales concerned with trifling objects such as galaxy clusters there is quite a lot of interesting variation that has produced stars, planets and people, but at larger scales the Universe seems much the same whatever direction you look in. If we look as far out as telescopes can see in one direction and then compare that to what we see in any other direction, there are no significant differences at all. In all directions we see very similar numbers of very similar-looking galaxies. However, in standard Big Bang cosmology, two portions of the Universe in different directions may never have had any contact with each other. For galaxies at the limits of observability from Earth, there has by definition only just been time since the origin of the Universe for their light to get to us. Two patches of sky in opposite directions, for example, are obviously twice as far from each other as they are from us, and even signals travelling at the speed of light cannot yet have crossed that distance. As far as each of these volumes of space is concerned, the other volume may as well not exist. But, if these volumes have never interacted, how can they look so similar? It's a bit like the completely unsubstantiated legend that there were tribes of Native Americans who spoke Welsh when European settlers first arrived. If true, the legend could only mean that there had been earlier contact between the two continents, since an independent evolution of Welsh is, to say the least, rather unlikely. In a similar way, completely independent evolution of identical cosmological structures is unlikely, and the implication is that there must have been earlier contact. But in standard Big Bang cosmology, this simply isn't possible. If we go back in time a few billion years, the light from those distant galaxies hasn't even had time to reach the Earth, let alone galaxies beyond our planet in the opposite direction. Quite simply, if these two patches of sky can't see each other today, they could not see each other in the past – and the further back you go towards the Big Bang, the bigger this problem gets.

Recent results from a spectacularly successful space mission have thrown this issue into even sharper relief. The Wilkinson Microwave Anisotropy Probe (WMAP) was launched on 30 June 2001, and after circling the Earth three times, it was sent towards the Moon whose gravity, in turn, slung WMAP out into interplanetary space. Six months later it reached its final destination, a spot 1.5 million kilometres from the Earth in the opposite direction to the Sun. This position, known as the L2 point, is a relatively stable place to put a spacecraft since it will stay put in that vicinity with relatively little adjustment using the spacecraft's own engines. The L2 point also has the advantage of being quite a long way from the Earth (which cuts down radio interference) while still being close enough to allow reasonably easy communication. Furthermore, the Sun, Moon and Earth all lie in the sunward direction, as seen from L2, and

so it is easy for the spacecraft to avoid looking at all three of
these over-bright objects. Once WMAP was on station, it began
an eight-year mission to take the temperature of the Universe.
Strictly speaking, WMAP actually measured differences in temperature between opposite points in the sky rather than measuring temperature directly, but that is a relatively minor technical
detail. WMAP's measurements were made using a device similar
to one that may have been stuck in your ear to take your own
temperature. In the ear-thermometer case, a detector measures
the amount of infra-red radiation emitted by your ear-drum
and uses this to determine body temperature. In WMAP's case,
its detectors picked up the microwaves produced by the much
colder background temperature of the Universe and found a
temperature of $-270.424°C$.

Some things in the Universe, stars for example, are much
warmer than this temperature, but most cosmic radiation does
not come from those relatively rare sources. The vast majority
of photons bouncing around the Universe are remnants from
the Big Bang, or more accurately, from a time 375,000 years
later when the Universe first became transparent. Prior to that
the Universe was so hot that electrons and atomic nuclei could
not combine to form atoms and light could not propagate
through this electrically charged plasma. Then, over a period
of just a few thousand years as the entire cosmos cooled below
about 3,000°C, atoms formed and photons were set free.
This, rather than the Big Bang itself, was the true 'let there
be light' moment. Since that time these photons have largely
moved unimpeded through the Universe. Cosmic expansion
has, however, cooled them so that they now indicate a temperature 1,000 times lower than that of the hot plasma from
which they originated. The microwaves we see today are the
red-shifted remnant of an ancient metamorphosis that changed
our Universe from being about as clear as mud to being far
more transparent than the purest of mountain spring water.

This remnant is called the cosmic microwave background and its existence, which was first demonstrated in the 1960s, is the single most important piece of evidence supporting the standard Big Bang model of the Universe, because it proves that the cosmos was once dense and hot. WMAP was measuring heat emitted more than 13 billion years ago and its mission was to measure more accurately than ever before how the intensity of this heat varies as we look in different directions. What WMAP produced was a heat-map of the sky showing fluctuations of just a few millionths of one degree. Places in the Universe that should never have been in contact with one another neverthe-less somehow manage to have almost identical temperatures. If it was just a few bits of the Universe that happened to be at the same temperature, this might be a coincidence, but it's not just a few bits. The whole of the early Universe was at more or less the same temperature and this implies thermal contact. Heat must have been able to move from hot parts to cooler parts so that temperatures equalised. Furthermore, the tiny remaining fluctuations that are present have exactly the right size to seed the later evolution of galaxies. Slightly cooler patches in the early Universe were slightly denser and therefore attracted, by gravity, material from their surroundings to make them denser still. Over hundreds of millions of years this process produced the galaxies and galaxy clusters that fill the Universe today. The large-scale uniformity of the galaxy distribution that I mentioned earlier can also be traced to the smoothness of the Universe's temperature a few hundred thousand years after its birth.

A few weeks after I wrote the preceding paragraphs, and while I was still working on the rest of this book, a new set of microwave background measurements from the European Space Agency's Planck mission were released. There are now three independent sets of space-based measurements of the micro-wave fluctuations. The first set came from the COBE mission, launched in 1989. Then came WMAP in 2001 as described above and, in 2013, we began to get the first results from Planck. With each successive mission the pictures have become a little sharper, a little more detailed and a little better at testing our understanding of the Universe. The cosmic architecture which emerges is a consistent one of a hot, early Universe whose temperature was extraordinarily uniform. There are some tiny fluctuations but these are themselves very uniform. We see similar deviations from precise uniformity whatever direction we look in and, as a result, the distribution and size of galaxies that formed later was also spectacularly uniform. The latest results from Planck do show one very minor but extraordinarily interesting feature not seen on the blurrier images from COBE and WMAP but I'll come back to that later. Based on the available data it is clear that, on the very largest scales, our Universe is extremely uniform with even the minor departures from perfect smoothness being surprisingly consistent whichever direction we look in. Given that even the early Universe was too spread out to have gained a uniform temperature by the usual route of heat moving from hot areas to cold ones, this needs explaining!

Fortunately, COBE, WMAP and Planck were not operating in a theoretical vacuum (no pun intended). An explanation for the Universe's smoothness and several other related mysteries had been proposed in the early 1980s by American cosmologist Alan Guth. Guth described the Big Bang as resulting from a process he called inflation, which gave rise to a young, hot and expanding Universe. According to Guth, inflation was a short period of enormous expansion in the very young Universe that took points that were microscopically close and moved them astronomically far apart in a tiny fraction of a second. As a result, points that were no longer in thermal contact after inflation had actually been close enough to equalise their temperatures when the process started. Essentially, a process of inflation stretches a young Universe so much that it is forced to be incredibly smooth. This expansion takes place at an unimaginably large rate that causes adjacent points to move apart much faster than the speed of light. This does not defy the normal laws of physics, however, because it is space itself that is expanding. There are no physical entities moving through space at these high velocities.

To explain the high degree of smoothness we actually observe in the Universe, inflation must have been so great that even points that are now a thousand times further away than the edge of the visible Universe must have been in thermal communication with us when inflation began. Detailed theories of inflation actually imply that this is a lower limit to the volume of that part of space that has the same properties as the visible Universe. Our Universe is therefore probably much more than a billion times larger than the observable Universe. However, and this is the fascinating new result from Planck, there are tiny departures from perfect uniformity in the microwave background data that may be the first indications of structure behind the veil of the 14 billion light-year horizon imposed by the finite speed of light. It's early days and there may yet be other, more mundane, explanations for these apparent deviations from a completely uniform Universe, but it would be hard to overstate how exciting these new results could be. We may be seeing the first signs of what the Universe looks like on a scale of trillions of light years.

But what physical process caused this massive expansion of the early Universe? We already have a mechanism for expanding a Universe: the cosmological constant produced by quantum mechanical fluctuations in the vacuum. But this is nowhere near strong enough to do the job. The accelerating expansion seen in today's Universe is a very gentle affair compared to that needed when the Universe was young. Guth suggested that the earlier phase of extremely rapid expansion resulted from a 'false vacuum'. I have, I hope, already convinced you that the vacuum should be thought of almost as a real physical object, an entity that has significant properties of its own. Now let's take that a step further. It is also possible that the vacuum can exist in different states. In the same way that water can be cold ice, warm liquid or hot vapour, the vacuum may exist in different forms with different energy content. The false vacuum is a state with much higher energy than that of the vacuum we see in our Universe today and, as a consequence, it has a much higher cosmological constant. False vacuum might therefore have been the cause of inflation in the very young Universe. This suggestion also provides an explanation for a very natural way for inflation to have ceased, since this high-energy form of vacuum will eventually condense into the less energetic true vacuum. That process of decay would have released vast amounts of heat to give rise to the hot Big Bang itself (there is a debate among cosmologists about whether we should think of inflation as leading to the Big Bang, as I've described it here, or think of the start of inflation as being 'The Big Bang' – but this is a purely semantic point and rather unimportant in my view).

Inflationary cosmology with its false vacuum and superluminary expansion seems like very wild speculation to those of us who are not cosmologists, but these ideas violate no known laws of physics and, more importantly, they make predictions about our Universe that fit the observations extraordinarily well. Not only does inflation resolve puzzles such as the smoothness of our Universe, it accurately predicts how the minuscule departures from a completely uniform temperature should look at different scales and it predicts how these small temperature differences gave rise eventually to stars and galaxies. Inflationary cosmology is very much mainstream cosmology these days. A multiverse, on the other hand, takes things to an even higher level of speculation and needs one final ingredient. Widely separated parts of the unimaginably enormous inflationary universe need to have different laws of physics. Thanks to physicists' almost obsessive search for simplicity, it begins to look as if such a thing is indeed possible.

I love Richard Dawkins's description of physics as the science of objects that are so simple you can actually use mathematics to describe them. There is more than a little truth in this, and as part of their eternal search for simplicity, physicists look for ways to unite things that seem to be very different: a falling apple is drawn towards the ground by the same force that keeps the Moon in its orbit, while a tsunami is described by the same equations that govern the flow of petrol vapour into a combustion chamber. In this spirit, physics has now been on a 150-year search to formulate what some call, with a touch of hubris, a 'theory of everything' in which all the different forces of nature emerge as different facets of a single force. This search has not been in vain. The laws governing electricity and magnetism were united in the 19th century, and an additional force, the weak nuclear force, was successfully added in the 1970s. There has also been significant progress with bringing the so-called strong nuclear force into the same theory. The final ingredient, gravity itself, is proving very tricky to incorporate but even here there are promising avenues of research. Despite the fact that this is still very much a work in progress, an important pattern has already emerged: the disparate laws governing the behaviour of objects from nuclei to molecules to galaxies start to look similar as we turn up the temperature. The expectation is that these laws had a very simple and completely unified form in the high-energy conditions that prevailed around the time of the Big Bang and that they then crystallised into their observed complexity as temperatures dropped. The key idea here is that the specific form into which the laws crystallised may not be fundamental; other laws may be equally possible. An often used analogy is that of a pencil balanced on its tip, which, when it falls, must do so in one specific direction. We would not expect it to fall in the same direction if the exercise was repeated. Similarly, the laws of physics may have fallen out in different ways in different parts of space. The massive expansion produced by inflation ensures that we are nowhere near different domains with different physics, but if these ideas are correct, those different domains are out there. These alien domains are so far away, and have such completely different physics, that it makes sense to think of them as completely different universes since we have no contact with them and they do things differently there.

It's probably worth mentioning that this is just one of many theories invented by cosmologists that give rise to multiple universes. Another related scenario assumes that inflation is eternal and merely breaks down locally to give rise to pocket universes of space that are no longer inflating, separated by unimaginably vast regions of ongoing inflation. Other scenarios examine whether the interior of black holes may constitute separate universes each with, in turn, their own black holes. Older theories have postulated an oscillating Universe that experiences repeated expansion, collapse and a 'Big Bounce'. A related concept in which multiple universes are separated in time is one where our Universe eventually becomes so dilute that a new Big Bang occurs spontaneously within it. From my point of view these different proposals are just details for the cosmologists to debate. The important point is that a multiverse is not as outlandish an idea as it might first seem and is one taken seriously by many experts in the field.

The picture that emerges from modern cosmology is that we live in a vastly larger, stranger and more diverse cosmos than the one envisaged just a few decades ago. This gives a stage on which the anthropic principle can perform its magic. Whether we regard the disparate regions of the cosmos with their varied laws and properties as truly separate universes or as simply distinct domains within a single Universe is merely a semantic point. Our laws of nature may be local by-laws whose strictures apply only in our small corner of existence. Under these conditions it is all but inevitable that we occupy a favoured location – one of the rare neighbourhoods where those by-laws allow the emergence of intelligent life. We do, in that sense, live in a lucky universe.

However, that does not necessarily mean that the view propounded in the rest of this book, that we live on a lucky planet, needs to be discarded. It is quite possible that both are true and that we live on a particularly favoured world even within our favoured portion of the cosmos. That brings us back to the 'unlucky planet' of Nemesis from the Prologue, one example of how not to build a highly habitable world. The difference between Earth and my imagined twin world was in their moons, and a look at how the size and distance of the Moon affect our climate reveals a real surprise. It is only by great good fortune that we have avoided a catastrophe that would have rendered our world incapable of supporting the complex and beautiful biosphere we enjoy.

12
Eclipse

I sat in a field and watched the Sun disappear. It was 11 August 1999 and I'd been waiting 30 years to see that day's total solar eclipse – ever since realising as a small boy that it would be the only one visible from Britain during my entire lifetime. I'd originally intended to head for Cornwall, where the 50-mile-wide eclipse track clipped the south-west corner of England, but then I had a better idea. The whole of Europe was bisected by the path of the Moon's shadow and clear skies with clear roads were much more likely on the mainland. The discovery that a friend's mother lived just off the eclipse track, in the German village of Schwabniederhofen, settled matters and I invited myself and five companions to stay for the week. I should add that our gracious hostess seemed delighted to have us.

However, as I sat in that field on the morning of the eclipse, Bavaria didn't seem such a clever idea after all. The solar disc had not disappeared behind the Moon, but, instead, had been obscured by a thick bank of cloud that had blown in from the west. Our only hope of seeing the eclipse lay to the east where the skies remained clear, and so we climbed back into our cars and weaved along German country lanes while trying to keep the Sun ahead of the advancing cloud-front and ourselves close to the centre of the eclipse path. Several hours later and with just seconds to go before totality, we pulled into a crowded field near the Austrian border and scrambled out of our vehicles. It was worth all the effort. The totally eclipsed Sun was an awe-inspiring sight with the jet-black silhouette of the Moon contrasted against the blue-white halo of the Sun's atmosphere and the pink, cloud-like solar prominences glowing within it. It wasn't just the eclipsed Sun that impressed. On the horizon in front of me, as I faced the Sun, I could see distant Alpine mountaintops glistening brightly in the sunshine beyond the southern edge of the Moon's shadow, while behind, eclipse-darkened clouds falsely threatened a thunderstorm of unimaginable violence. After one minute of the two-minute eclipse, the weather front we'd been outrunning for the previous two hours finally claimed victory as its clouds covered the still eclipsed Sun. No matter, we'd seen what we'd come to see and our motorised pursuit of an ever-shrinking patch of blue sky is now as precious a memory to me as that of the eclipse itself. That was my first-ever eclipse but I hope it won't be my last; especially as I forgot to give my wife a go with the binoculars. Perhaps I should take a spare pair next time.

Eclipses are undoubtedly among the most beautiful of nature's spectacles but many people, including me, are also struck by their sheer implausibility – the fact that eclipses occur only because the Sun and Moon happen to look almost exactly the same size. The Sun is really 400 times wider and 400 times further away but that doesn't lessen the impression that something very odd is going on. Nagging feelings are not usually a good basis for solid science, and most experts dismiss eclipses as a meaningless, although scientifically useful, coincidence. But there is another way to look at it. Perhaps a Moon big enough to obscure the Sun might make our planet better suited for life than it would otherwise be, and if this is the case, planets that by chance happen to experience eclipses are more likely to give rise to intelligent life than planets that do not. At first this seems an insane suggestion. How could a complex and beautiful biosphere like the Earth's possibly be more likely on worlds that have eclipses, phenomena that occur, at any given place on the Earth's surface, only two or three times in a thousand years? However, the bigger a moon looks, the stronger the tides it generates, and, with tides, we do have something that directly affects life. A moon is not needed for tides, since the Sun also produces them, but tides are significantly increased in strength if a planet has a sizeable moon too. Tidal forces are doubled on a planet with a moon that looks 80 per cent as large as the Sun and tripled if, as happens on Earth, the Sun and Moon appear equal in size. Critically, as I discussed when looking at the causes of ice ages earlier in this book, tidal forces indirectly affect our climate by controlling how fast and how much the tilt of the Earth's axis changes through time. Eclipses could therefore indirectly contribute to the life-friendliness of our planet if worlds that happen to have big-looking moons, and therefore large tides, also have better weather.

This idea that there may be a direct link between our large Moon and the habitability of our world was irresistible to me. I have always been fascinated by our satellite. Whether by day or by night, whatever phase she shows and whether she sits red and bloated on the horizon or shines her silvery light from high in the sky, the sight of our Moon never fails to lift my spirits. She has also played a concrete role in shaping my career. Like many people of my age, I developed a lifelong interest in science and technology as a direct result of the Apollo programme that took men to the Moon for the first (and so far only) time in the late 1960s and early 70s. I vividly remember sneaking out of bed, as a small boy, to listen to the live broadcast from Apollo 11 as it landed. Forty-four years later I retain a crystal clear memory of the transistor radio, with its plastic red and silver buttons and silver speaker grille, which I pressed against my ear to hear Neil Armstrong announce from the lunar surface that 'the Eagle has landed'. Later, as a teenager, I built my own telescopes in the back garden and had the love–hate relationship with the Moon that many stargazers have. It's a beautiful alien world and the only one whose surface can be seen clearly through small telescopes, but on the other hand, it always seems to be

full and bright when you want to catch a fleeting glimpse of a
faint comet or count the shooting stars in a meteor shower. As
I grew older my scientific career drifted away from astronomy
and into the related and equally fascinating realm of geology,
and so I stopped looking up and started to look down. Decades
later still, after reading Barrow and Tipler's *The Cosmological
Anthropic Principle*, I started to think about whether the Earth
might have unusual, life-enabling characteristics. It was clear
to me that looking at the Moon was the best way to test that
intriguing idea. Was our large Moon an example of a rare property that helps make our world highly habitable? Is it part of
what makes the Earth special?

It may seem surprising that the Moon could provide the
best evidence of the Earth's life-friendliness when other factors, such as biological evolution, have had a much more direct
and significant impact on our planet's developing environment.
There are several reasons why the Moon tells a more convincing story of our good fortune than many other, apparently
more promising, facets of our world. For a start, the behaviour of the Earth–Moon system is a reasonably well understood
one, controlled by the relatively simple equations of celestial
mechanics. I say 'relatively simple', because the details are still
a bit of a nightmare. Isaac Newton himself complained that
thinking about the motions of the Moon made his head ache!
Nevertheless, unlike climate evolution or the evolution of animals and plants, the changing behaviour of our satellite through
time can be mathematically modelled with reasonable precision.
In the jargon of these things, modelling of the Earth–Moon
system is 'tractable', and unlike poor old Newton who only
had his towering genius to rely on, we at least have electronic
computers to help with the number crunching. There is yet
another reason why the Moon is a good target for study when
investigating the life-friendliness of our world. If we have an
Earth–Moon system whose characteristics happen to make the

Earth particularly habitable, then this can be used to distinguish between the three scientifically respectable explanations for our good fortune in living on a planet as benign as Earth. Let me recap these in the light of the material covered in the preceding chapters.

Firstly, it could be that highly habitable worlds are fairly common because mechanisms, such as volcanic weathering and limestone deposition or the possible Gaian properties of a complex biosphere, emerge naturally from the physics of a life-friendly universe. Thus, we shouldn't be surprised by the Earth's suitability for life because the laws of the Universe guarantee the existence of many such worlds. Alternatively, it may be that life is extraordinarily adaptable and will thrive under a wide range of conditions. Thus, we shouldn't be surprised that the Earth fits life because, in fact, life has adapted to fit the Earth. Finally, perhaps well-regulated planets occur only very rarely and purely by chance, but because the huge size of our Universe allows many attempts at constructing even the most peculiar of worlds, such places are still inevitable. We shouldn't then be surprised by our good fortune in living on one of these oddballs, because we must find ourselves inhabiting one of the lucky worlds that had the billions of years of good conditions necessary to produce a complex biosphere and, ultimately, intelligent observers.

These three possibilities are not mutually exclusive. The evidence that our Universe is surprisingly life-friendly and that life is remarkably adaptable is compelling. But I don't think that is the whole story. Luck must play an important role, too. For one thing, it's not necessary for highly habitable worlds to be common in the Universe from the point of view of explaining our existence. Even if only one planet in a trillion is habitable there will still be an unimaginably large number of such worlds in the Universe. I don't think there is any evidence at all to suggest that life-friendly worlds are substantially more common than this, and so, in my view, imaginatively populating our small corner of one galaxy with hundreds of advanced civilisations is just wishful thinking. The scientifically conservative position should be that life is rare and intelligence even more so.

However, more evidence is needed to substantiate this pessimistic view, and that's where the Moon comes in. Only anthropic selection, the unavoidable observational bias that occurs because we must live on a habitable world even if that requires some really odd features, can explain a benign Moon. If properties of the Moon, such as its size and distance, turn out to be fine-tuned to make the Earth more habitable, this cannot be the result of feedback or of Gaia; there are no mechanisms to move the Moon or modify its size to compensate for poor environmental conditions at the Earth's surface. Lunar fine-tuning also cannot be the result of natural selection, because the Moon is obviously not a living organism. An Earth–Moon system with particularly good biosphere-enhancing properties cannot even be the result of living in a life-friendly universe, because moons with other sizes or separations are entirely possible. We can see this just by looking around our own solar system. If the Earth–Moon system is fine-tuned for life, with the size of the Moon, for example, being 'just right', then at least one of Earth's chance peculiarities is a necessary precondition for the emergence of intelligent life.

With this analysis I had found a good reason for me, as a geoscientist, to investigate the Moon, and it felt like coming home. I'm probably a bit of a Moon bore, if truth be told, and I'm often astonished by friends' misconceptions. Many people, for example, believe that the Moon comes out only at night when in reality it's up in the daytime just as much as it is at night. I'm also easily irritated by minor Moon mistakes in novels. One book I've been trying to read recently has a half-moon rising at twilight. This is simply not possible. If it rises around sunset, the Moon must be nearly opposite the Sun and will therefore be full. I've been struggling to take the rest of the book seriously ever since, which is a pity as the mistake happens on page five. Television is often not much better. The Moon seen through the bedroom window of Peppa Pig, my daughter's favourite cartoon character, is always a crescent illuminated on the left-hand side. But a crescent Moon in the evening is lit from the west, which will be on the right as seen from the northern hemisphere. Peppa therefore either lives in the southern hemisphere, despite the Pig family's strong English accents, or she has an unreasonably late bedtime for such a young pig. These are trivial issues, of course, and it is childish of me to be annoyed by them, but they do illustrate how rarely modern people look at the sky. Our ancestors would never have made such mistakes.

Recently, I've found myself a whole new lunar misconception to correct, and this is one propagated by professional astronomers rather than cartoonists or novelists. It is widely believed by experts that the strong tidal forces generated by our large Moon help to stabilise the Earth's axis. This fits nicely with the 'eclipse coincidence' with which I began this chapter. If a large Moon is needed to prevent the climatic chaos that would result from a wildly tumbling Earth, then it is no longer surprising that we inhabit a world where the Moon is both large enough to stabilise our axis and, as an entirely fortuitous side-effect, large enough to cause eclipses. However, the idea that large moons stabilise the axes of their parent worlds is simply wrong. The next chapter looks at why.

The Dark Side of the Moon

Like many people, I am fascinated by counterfactual histories – histories that imagine what would have happened had events in the past taken place differently. There is a popular genre of writing that plays such games with human history. What would have happened if Suleiman the Magnificent had captured Vienna in 1529 so that Islam became established at the heart of Europe? What would have happened if the Viking settlements in North America had not died out 400 years before Columbus arrived in the New World? What would have happened if Danton, rather than Robespierre, had won the battle for dominance of the French Revolution? We can play similar games with the Earth–Moon system and speculate about what would have happened had things been just a little different when the Earth and the Moon formed. In particular, what would have happened had our Moon been a different size. I should start by explaining how the Moon formed in a little more detail than I gave in my rapid tour of Earth history back in Chapter 4.

The Moon's origin has long been a topic of debate, with some astronomers believing that it was ensnared by our world when it happened to pass nearby, and others suggesting that it is a piece of the Earth that broke off. These can be summarised as the 'capture theory' and the 'fission theory'. Fission was the favoured theory of the Moon's origin for many years and it was believed that the Pacific Ocean was the scar left behind by the departing Moon. However, as we now know, the Pacific is far too young to have had anything to do with lunar origins, and in addition, it is hard to see why the Moon would suddenly have broken off in this way. What changed to make that particular lump of rock want to leave? The alternative capture theory therefore came to be preferred by many experts, and the debate between the two camps was, as often happens at the frontiers of science, a very heated one.

In the end, however, elements of both ideas turned out to be correct. The argument was settled, to most people's satisfaction, by analysis of lunar rocks brought back to Earth by the Apollo programme. These showed the Moon's composition to be almost identical to that of the Earth's mantle, and the similarities were too strong to be explained as resulting simply from the Moon and Earth forming in the same part of the solar system. Analysis of lunar rocks also showed that the Moon's surface solidified between 50 and 150 million years after the solar system's origin. This evidence therefore favoured the idea that the Moon somehow broke off from the young Earth, but there was still a problem with explaining how this could possibly happen. As I described in Chapter 4, the now widely accepted explanation is that the Moon was forged by an impact between two worlds at a time when the solar system was still young but no longer quite brand-new. The impact knocked a substantial volume of the larger planet's mantle into space and some of this settled into orbit around that world to give it rings to rival those of Saturn. However, this orbiting rubble was too far out to suffer the gravitational disruption that helps maintain Saturn's rings, and so the temporary rocky girdle rapidly amalgamated into a moon within just a few thousand years of the impact. The result was the only double-planet in the solar system: the Earth and its Moon.

There are still a few problems with this account. In particular, when the impact is modelled by computer it's easy to get a Moon that's the right size and it's easy to get a Moon with the right composition, but it's proving difficult to find an impact that reproduces both properties. Nevertheless, most experts are convinced that this theory for the Moon's origin is essentially correct, even though we still need to find an impact scenario that fully accounts for all the facts.

The resulting Moon was ten times closer to us than it is today, and the world it orbited had been spun up by the enormous Moon-forming collision to give a day a little over five hours long. Now, 4.5 billion years later, our satellite has drifted out by 350,000 kilometres while the Earth's day has grown to 24 hours. The engines of change were the tides. As the Earth rotated under the tidal bulges raised by the Moon in our oceans, the friction of the resulting currents across the sea floor gradually slowed the Earth's spin and our days became longer. This drag against a rapidly rotating Earth also pulled the locations of the highest tides forward so that, instead of lying directly under the Moon, they were a little ahead. High tides, even today, tend to happen a little after the Moon is at its highest point in the sky, although this simple picture is distorted substantially by the effects of water sloshing about in the complexly-shaped seas and oceans of our world. Nevertheless, the weak gravitational attraction of these offset bulges gently tugs the Moon forward and, as a consequence, the Moon has very gradually moved into a higher orbit. The detailed mathematics of all this was first worked out by George Darwin, Charles Darwin's second son, who was born in 1845 and became one of the most celebrated astronomers of his generation. Darwin was a supporter of the fission theory for the Moon's origin and one of his many lasting contributions to astronomy was to show how the Earth–Moon system would have evolved after it broke apart when our world was young. Fortunately, his maths works just as well for the modern impact explanation.

Thanks to laser-reflectors left on the Moon by the Apollo astronauts we know that the Moon is still drifting away from the Earth, by about 4 centimetres a year, and that our days are still getting longer, by about twenty seconds every million years. However, according to Darwin's mathematics, the Moon receded about a thousand times more quickly when it was young and drifted out by 100,000 kilometres within the first few million years. This was a time before the Earth had oceans, but tides affected the solid Earth as well as the early Earth's oceans of molten magma. Tidal drag therefore began its transformation of the Earth–Moon system even before our world had seas.

Given this history for the origin and evolution of our satellite, it's a simple matter to come up with counter-factual histories for the Earth–Moon system. The planet that hit our world might have had a different size, or the collision could have been closer to head-on or more of a glancing blow. The impact could have happened with a different closing velocity too, depending on whether the two planets were orbiting the Sun at nearly the same speed and in nearly the same direction when they met. Any such differences between these hypothetical collisions and the one that actually occurred would have produced a Moon with a different size and left the Earth spinning at a slower or faster rate. The collision may also have played an important role in setting the obliquity of our planet and so this, too, might have been different to the 23 degrees we actually see.

There is also no reason why the tidal drag produced on these alternative Earths would necessarily be the same as that which actually occurred. The average drag on the real Earth over the last 4.5 billion years is easy to calculate, because it has to be just the right size to move the Moon from being 30,000 kilometres away, when our world was young, out to 384,000 kilometres today. Remarkably, this average drag turns out to be only one third of the size needed to explain the current 4 centimetres per year recession rate. So tidal drag must have changed significantly through time and seems to be exceptionally strong at present. The strength of the drag depends on the size and shape of the Earth's ocean basins, and these altered massively as the continents slowly grew and drifted across the Earth's surface. Tidal drag strength also depends on how fast the Earth is rotating, and this slows through time as we've already seen. Computer modelling suggests that tides could be particularly large for an Earth that takes 20 to 30 hours to rotate, and this may well explain why tidal drag is so strong at present. Our alternative Earths are unlikely to have exactly the same continents as our own world and will be spinning at different rates, and so the average strength of tidal drag will inevitably vary a little as we move from one counter-factual Earth to another. Thus, even if the initial spin rate and lunar size had been identical to our own, the Earth–Moon separation and Earth day length on my counter-factual worlds will differ from ours after 4.5 billion years of evolution.

Counter-factual Earths with different-sized Moons at different distances and with different-length days must actually exist. The staggeringly enormous number of planets in the Universe means that the Earth has many near-twins that only differ significantly from the Earth because their moon-forming collisions were not quite the same as our own. How will this make those worlds differ from ours? When I began this work I believed, in common with everyone else in this scientific field, that a large Moon helps stabilise the Earth's axis. My expectation was therefore that 'Earths' that happen to have smaller moons would have chaotically fluctuating obliquities and be unpromising locations for the maintenance of complex biospheres. Furthermore, I thought that to provide the necessary stabilisation a moon would need to look big enough to obscure the Sun, thus neatly explaining the eclipse coincidence. I was completely wrong!

The first surprise was that all these potential changes wouldn't greatly alter the apparent size of the Moon. A larger Moon would have raised bigger tides and, as a result, receded faster from the Earth to end up at the present day looking much the same size as the Moon we actually have. A Moon twice as massive as the real one would, after 4.5 billion years of drifting away from the Earth, now lie 44,000 kilometres further away than our Moon and would look only 8 per cent bigger; a difference so small that you would find it hard to spot by eye. Thus, over a wide range of sizes, a Moon produced by an impact with an Earth-like world ends up looking much the same size as the Sun, since larger Moon-sizes are almost cancelled out by greater Earth–Moon distances. This conclusion isn't altered much even if the tidal drag and initial spin rates are changed substantially. The apparent size of a 4.5 billion year-old Moon is relatively insensitive to the initial conditions when it formed and relatively insensitive to the exact strength of tidal drag in the oceans. It's not even particularly sensitive to the age; after a few hundred million years a moon's recession rate slows down and, thereafter, its distance changes only very slowly. Eclipses still need a coincidence but it's not such an extraordinary one after all, since almost any Earth–Moon-like system that's several billion years old will put you in the right ball-park.

There is, in fact, only one thing that will change this picture. The Earth must have been spinning reasonably fast immediately after the collision. If this is not the case then the Earth–Moon system runs out of steam before the Moon gets far out. If the Earth is initially rotating quite slowly it can only spin down to the point where its day has the same length as the duration of the Moon's orbit. The Earth is then tidally locked to a Moon that is now in a geosynchronous orbit; an orbit where the Moon is permanently stationed above a fixed point on the rotating Earth's surface. The tides raised by the Moon are then no longer moving with respect to the Earth and tidal drag ceases. The Earth and the Moon will thereafter turn the same face to one another for the rest of time and their separation too will become stuck at the moment of tidal locking. The resulting Earth would be a very different world to the one we inhabit. The tidal locking would, literally, make each day last for a month.

The resulting climatic effects would be severe. There would no longer be such a strong temperature difference between equator and pole; rapidly rotating worlds have colder polar regions because spinning deflects warm currents of air or water travelling out from the equator. This Coriolis effect, as it is called, produces the complex pattern of rotating weather systems that dominate the mid-latitudes of our atmosphere. As a result of this Coriolis deflection, transport of heat away from the tropics is less efficient than it would be on a more slowly spinning Earth and, as a consequence, our poles are colder than they would otherwise be. In contrast, an Earth with a long day has relatively warm poles and, instead of a temperature contrast between poles and equator, such a world would have a strong contrast between the day side and the night side. Overall, the weather patterns driven by this very alien temperature distribution would be unrecognisable and the habitability consequences very hard to determine. In the rest of this chapter, however, I will ignore such worlds and restrict my attention to counter-factual planets that are more like the Earth we know and love, a planet with a large Moon but with no tidal locking of the bigger partner just yet.

As I've already said, eclipses don't need much of a coincidence on non-tidally-locked worlds. Even more surprisingly, a large moon does not stabilise the axis of such planets. This conclusion contradicts so much received wisdom that I really need to explain it in depth.

Unimpeachable work by world-class scientists has shown that, if the Moon were to suddenly disappear, Earth's axis would become unstable. I have no argument with this conclusion at all. However, the implication has been widely drawn that the Moon therefore stabilises our axis. This may seem logical, but it's the right answer to the wrong question. We shouldn't ask: 'What would happen if we magically removed the Moon today?' Instead, we should ask: 'What would have happened if we had had a larger Moon from the beginning, 4.5 billion years ago?' Surprisingly, the second question gives the opposite answer to the first. As I'll discuss below, planets possessing a moon natur- ally evolve towards a state in which their axes are unstable, and this happens more quickly if the moon is large. Large moons therefore cause, rather than prevent, axial instability. But why does instability happen at all?

The spin axis of a planet doesn't always wobble in the nice, well-behaved way I described earlier in this book. Sometimes the axis wobbles instead in an uncontrolled and chaotic manner and this will happen to counter-factual Earths if they happen to have unsuitable moons. As you'll recall from an earlier chapter, the Earth precesses gently on its axis while, at the same time, the orbits of all the planets in our solar system experience changes to their orientation and shape. You may also recall that the time taken for the Earth to wobble depends on how fast she is spin- ning and how strong the tides are. These will be different on counter-factual Earths and so these alternative worlds will not precess every 26,000 years, as the Earth does, but instead take a longer or shorter time. As long as the resulting axial precession of these worlds happens at a different speed to the wobbling of the planetary orbits there is no problem, but if the frequencies happen to match, disaster results. This is due to resonance, a term that perhaps needs a little more explanation.

Imagine pushing a child on a swing. You would naturally push just after she has passed her point of closest approach, and so you would give her a periodic shove at precisely the same frequency as that of the swing. This is an efficient way to do the job and you can keep the child happy with minimum effort. However, if you weren't carefully watching what you were doing and frequently pushed the swing at slightly the wrong moment, the result would not be so good. Sometimes you would be speeding the swing up and sometimes you would be slowing it down, and, overall, the child's ride would be very poor. When you push with the right frequency the result is a large-amplitude swing, and this strong result for relatively small effort is described as resonance.

Resonance is not always such fun. It can cause catastrophe. A good example of this is the rather odd effect, sometimes seen in earthquakes, where small buildings fall down while taller ones remain standing. If shorter buildings happen to vibrate at similar frequencies to shaking from that particular quake then they resonate and, as a consequence, wobble so violently that they collapse. The best documented case is from a magnitude 8.1 earthquake that hit Mexico City in September 1985, killing at least 10,000 people. The shaking had a predominant period of two seconds and most of the buildings that collapsed were between five and fifteen storeys high. Tall buildings take substantially longer than two seconds to oscillate back and forth, while shorter ones vibrate significantly faster, and so both of these were relatively safe. However, the intermediate-height structures resonated with the two-second shaking and shook so violently that many of them collapsed with terrible loss of life.

Catastrophic resonance can also happen to planets. The gravitational influence of any other planet on the Earth is absolutely tiny, but if it happens to give us a very slight nudge at exactly the same frequency as the Earth is wobbling anyway, then, just as with pushing a child on a swing or shaking a building in an earthquake, the effect builds up into something very significant. Resonance between planets therefore throws a large spanner into their normally clockwork-like behaviour and, when this happens, the result is disaster. Computer models show that if the Earth experienced such resonance, the orientation of our axis would change by up to 50 degrees over a few million years and the resultant continuously changing climate would make living conditions very unpleasant for most organisms. Mars actually does resonate with the solar-system perturbations and, as a consequence, experiences periodic climate change on a scale that makes Earth's ice ages look about as serious as an English summer shower.

Fortunately for us, our planet does not share Mars's problem; at least not yet! The Earth currently precesses once every 26,000 years while planetary orbits oscillate with periods of 50,000 years and upwards. However, as the Moon continues to recede and the Earth's day continues to lengthen, this will change. The Earth's precession speed will be reduced by the smaller equatorial bulge of a less rapidly spinning world acted on by the smaller tidal forces of a more distant Moon. The Earth's precession rate therefore will steadily decrease over time and, in about 1.5 billion years, resonance will occur. From that moment on, the Earth will have an unstable spin axis.

This is the inevitable fate of any Earth-like planet with a moon. The planet's spin will slow and the moon recede until, eventually, resonance with the other planets produces a chaotically tumbling planet. If the planet has a large moon, this drift towards chaos will happen more quickly. That is why giving the Earth a larger moon 4.5 billion years ago would have given us a chaotic obliquity today.

The contradiction between the story usually told – that our large Moon stabilises Earth's axis – and the correct story occurs because a large moon is a two-edged sword. A moon increases axial stability by increasing tidal forces but it also increases the speed with which an initially stable planet races towards catastrophe as its rotation slows and the moon recedes. This reminds me of a race that took place in 2012 between the fourth in line to the British crown, Prince Harry, and the world champion sprinter, Usain Bolt. The occasion was a photo-opportunity at a sports stadium during the prince's visit to Jamaica. As a joke, the royal visitor beat the world's fastest man by running down the track before Usain Bolt even knew that a race was on. However, Prince Harry is a fit young man so he could probably give Usain Bolt a good run for his money in a 100-metre sprint; but only if he had a 20-metre head start. It would then be a pretty close finish. Would Usain Bolt's greater speed compensate for the extra distance he'd have to cover? Planets racing towards instability as their rotation rates drop are in a similar position. Which planet would win the race for instability: a large-mooned planet that moves rapidly towards this finish line from a long way off, or a more slowly evolving small-mooned system that starts from a point already close to instability? Mathematical modelling gives an unequivocal answer: large-mooned planets (and Usain Bolt) win the race every time! An initially stable planet with a large moon will become unstable long before an initially stable planet with a small moon.

An unstable axis because of too large a moon was the fate I imagined for Nemesis in my Prologue. The Earth has been more fortunate than Nemesis because our Moon is smaller. The bigger moon of Nemesis raised marginally larger tides than those on Earth and these slowed its rotation slightly faster and moved its moon out a little further than our own. The resulting drop in precession took Nemesis into the zone of instability just as it got to the age when dinosaurs ruled the Earth (and dragons ruled on Nemesis). Large moons do not stabilise planetary axes, in fact they do the exact opposite.

So, eclipses don't need much of a coincidence and our large Moon doesn't stabilise our axis. Has my quest to use the Moon to demonstrate the specialness of our planet therefore failed? Actually, no! When I first saw the results of my calculations I was struck by an extraordinary coincidence. Our large Moon is almost too big. If the Moon's radius had been just 10 kilometres bigger and the early Earth day just ten minutes longer, the Earth's axis would be about to become unstable today. Keep the Moon and initial day length that we had, and instead increase the average tidal drag by just a few per cent, and the same thing happens – the modern Earth would be an unstable world that could not sustain us. Of course, this near-instability could just be a coincidence. After all, provided the Earth's axis is stable it doesn't really matter how close it is to being unstable. Being almost unstable is a bit like being almost pregnant; if you don't want a baby, 'almost pregnant' is good enough. However, it was a very close call. Less than 1 per cent of stable, counter-factual Earth–Moon systems are closer to instability than the true Earth–Moon system. I think this makes the coincidence sufficiently unlikely that it's worth considering possible explanations for it. I may have one.

Near-instability is exactly what we would expect if big moons are helpful to life for an additional reason unconnected with axial stability. If large moon sizes really do favour complex life, but there is also an axial stability constraint on the biggest size allowed, the result should be a moon that has almost exactly the largest diameter permitted. The combination of an upper limit with pressure to be large naturally puts you close to the maximum allowed. The effect is similar to one you've almost certainly seen on fast roads. The average speed on British motorways, for example, is close to 70 miles per hour because that's the UK speed limit and everyone's in a hurry. Similarly, if intelligent life is more likely to emerge on planets with large moons, as long as they're not so large that the planet's axis becomes unstable, then intelligent observers will tend to find themselves staring up at a moon that is almost, but not quite, too big for comfort.

So, why might bigger moons be better as long as they are not too big? Here I have to be speculative, although I hope my guesses are at least reasonable. The key properties that result from having a Moon that is almost, but not quite, too large are that the Earth's axis precesses relatively slowly (almost slowly enough to be unstable) and that the Earth has a relatively long day (the only way to give a slowly spinning planet a stable axis is to add in a large moon). Both of these factors affect the intensity and frequency of ice ages, and my suspicion is that a moon almost large enough to cause axial instability allowed our planet to have relatively mild and infrequent ice ages. For a start, the 41,000-year obliquity variation that has driven much of the waxing and waning of ice sheets over the last 2.5 million years would be quicker if the Earth's axial precession was faster and therefore further from the critical value at which axes become unstable. Thus, on a planet further from instability than ours, the ebb and flow of ice ages will be more frequent. Furthermore, because of the Coriolis effects I mentioned a little earlier, an Earth with a shorter day would have more extensive ice caps. It would therefore reflect more heat into space, making the whole planet cooler on average. In a nutshell, rapidly spinning 'Earths' are more prone to ice ages.

It seems that the precise size of the Moon and the precise length of our day are fine-tuned after all. Had the Earth's day after the Moon-forming collision been a few minutes longer or had the Moon been a few kilometres larger, the Earth would now precess so slowly that there would be resonance with the wobbling orbits of the other planets and our world would have an unstable axis. If, on the other hand, the day had been shorter or the Moon smaller, modern Earth would precess faster and spin more quickly, giving us more frequent and severe ice ages. The true Earth–Moon system sits in a sweet spot between the life-destroying fates of frequent, severe glaciation or climatic chaos. There are probably billions of Earth-like worlds spread throughout the Universe but most of them will not have been gifted with the moon they deserve, a moon with just the right properties to ensure minimal astronomical interference with climate. A few worlds will, however, have been lucky and will have a moon, a day length, a tidal history and an obliquity that sustain clement conditions for long enough to allow the rise of intelligent observers. This is the best example I know of the observational bias I introduced at the start of this book. We can only possibly observe a planet whose properties allow our existence, and the Earth–Moon system's characteristics are a particularly clear example of the oddities this can produce.

Gaia or Goldilocks?

For its first half-billion years the Earth was a hot, dangerous and sterile world. This was the era of massive bombardment, the time when planets were created. Then the impacts stopped, the Earth cooled and life began. Ever since, our climate has been suitable for life, with our seas neither boiling away nor completely freezing. Four billion years of good weather is so surprising that it needs explaining, and that is what this book has been about. In my view, there are really only two games in town: either the Earth system itself maintains a rough equilibrium or our planet has been extraordinarily fortunate. It's either Gaia or Goldilocks.

I hope it's obvious by now that my preference is for Goldilocks, even though Gaian processes (broadly interpreted to include uncontroversial ideas such as volcanic weathering) must have played a substantial role too. However, even if I have been a little negative at times, there's a lot that I admire about the Gaia hypothesis (even when interpreted narrowly). It is an ambitious theory, but there's nothing wrong with ambition and James Lovelock's big idea has played a significant role in stimulating the modern view of the world as a single system composed of many interacting and inter-related geological, biological and climatic mechanisms. I particularly admire its name.

'Gaia' was suggested by the novelist William Golding in an act of superb public relations. Gaia was an ancient Greek goddess whose name is sometimes translated as Mother Earth. She was the second of the gods to come into existence, following Kaos, and they were both part of an older generation that ruled before Gaia's grandson, Zeus, overthrew his father, Kronos. Despite this regime change, Gaia retained her position of influence by adapting superbly to the changed circumstances. However, among the new gods on Olympus was Phaethon, who stole the Sun-chariot from his father, Helios, and drove it too close to Gaia. Only the intervention of Zeus, who struck Phaethon dead with a thunderbolt, prevented the complete destruction of Mother Earth.

As we've seen, modern science tells a remarkably similar story of a highly resilient Earth surviving dramatic changes in circumstances but destined to be scorched by the Sun. Our planet has gone through four main phases during its long history: first an abiotic time of massive bombardment, followed by the bacterial empire of anoxic stability, then the rise of an oxygen-rich atmosphere, leading finally to our own era of multi-celled plants, fungi and animals. At every stage Earth has serenely fitted in with each new regime as if nothing untoward were happening. However, in less than a billion years' time the Sun will become too warm for life on Earth to continue. At the very least our planet will be severely overheated and, after a few more billion years, it may even become engulfed by the Sun as it evolves into a red giant. It's unlikely that Zeus, or anyone else, will save Earth from terminal damage.

So, naming the idea that we live on an inherently benevolent and adaptable planet the 'Gaia hypothesis' was very appropriate. With tongue firmly in cheek I'm tempted to call the alternative idea, that we were just plain lucky, the Nemesis hypothesis. As I mentioned in the Prologue, Nemesis was the Greek goddess of undeserved good fortune whose main function was to restore balance by dealing out retribution to those who had got away with too much for too long. But Nemesis was also the mother of Helen of Troy. Nemesis therefore seems to me to be a highly apt designation, because if anthropic ideas are true, unearned luck has placed us on a world of extraordinary beauty.

Nemesis will exact her price in the long term, but fortunately not for hundreds of millions of years. Nevertheless, within a billion years even a zero level of carbon dioxide in the atmosphere will not be low enough to stop the ever-warming Sun from overheating our globe. Even worse, if the weathering-mediated feedback I discussed in Chapter 5 is sustained, then the carbon dioxide level in our atmosphere will continue to fall and, well before a billion years from now, it will become so low that plants will be unable to grow. Plants have begun to adapt to the very low levels of carbon dioxide we already have, but there is a limit to how far such adaptation can go. No conceivable plant could grow if there were no carbon dioxide at all in the atmosphere. Present models suggest that plant life might be able to limp along for another half-billion years or so but that's about it. If, on the other hand, this negative feedback is unable to completely compensate for warming, carbon dioxide will not fall fast enough and our planet will become uncomfortably warm within a few hundred million years. Either way, the distant future for plant and animal life is bleak.

Even if our luck holds and some unexpected combination of processes succeeds in keeping temperatures down and carbon dioxide levels up, the Earth's axis will become chaotically unstable in 1.5 billion years and this will certainly produce some very unpleasant conditions. So the idea frequently discussed in astronomy textbooks and science fiction stories, that the Earth will remain habitable for another 5 billion years, is hopelessly optimistic. If our distant descendants are going to survive they will need to take control and shield the planet from an increasingly hostile Sun. Indeed, if the thrust of this book is right and our 4 billion years of good weather has to a significant extent resulted from good luck rather than natural control mechanisms, we may have to take charge much sooner.

Still, from a human perspective that's all in the extremely distant future. Are there reasons why the ideas discussed in this book should matter in the here and now? One reason for caring about anthropic thinking is that it may be a good, and rather neglected, tool for expanding our knowledge about the world around us. People have been thinking about anthropic ideas for a very long time but perhaps the earliest example of its use in a scientific debate was the late 19th-century controversy concerning the age of the Earth. The anthropic argument was simply that the Earth must be old enough to allow the evolution of human beings, that is, a young Earth is inconsistent with the observation that we exist. Although they wouldn't have put it quite that way at the time, this argument led both geologists and biologists to conclude that the Earth must be hundreds of millions of years old. Physicists scoffed at this notion because their calculations showed that the Earth's interior was too warm to be that old and, moreover, it was not possible for the Sun to shine for so long. Given this situation, Thomas Chamberlain, professor of geology at the University of Chicago, concluded in 1899 that 'atoms [have] locked up in them energies of the first order of magnitude'. He also speculated that conditions at the centre of the Sun may allow this energy to be released. Chamberlain's suggestion was years ahead of its time, decades before the discovery of nuclear fission gave a mechanism for keeping the Earth warm or the discovery of nuclear fusion showed how the Sun could shine for such an immense period of time. Nuclear power was invented by a geologist! Chamberlain's prediction that new sources of energy would be found hidden in the Sun is one of the best examples I know of a successful prediction based on anthropic selection.

This was a rare success for anthropic reasoning, although there have been many uses of such ideas to explain already recognised facts. For example, it was known by the end of the 19th century that the Earth must be sufficiently large for its gravitational field to retain an atmosphere, and this explains, in retrospect, why the Earth is an unusually large rocky body.

Similar arguments have been proposed to explain the location and mass of Jupiter, the existence of plate tectonics, the particular location of the Sun in the galaxy, and a great many other features of our planet and planetary system.

However, these 'just so' stories remain highly speculative and, in any case, do not provide the strong support for anthropic ideas that would result if new and successful predictions were made ahead of observations. Now is the best of times to make such predictions. Thanks to ongoing and planned exoplanet characterisation projects, we will have a great deal of new information concerning thousands of planetary systems within a few decades. We have an unparalleled opportunity to make anthropic predictions that can be tested within our own lifetimes. I'm going to finish *Lucky Planet* by sticking my neck out to make some.

The conclusion I draw from all that I have discussed in this book is that, statistically, we are most likely to be living on a second-best world in a second-best universe. This is a universe where planets as well suited to life as the Earth are rare but not vanishingly rare and where even simple life is a tough trick to pull off. We therefore live in a moderately habitable, rather than highly habitable, universe and this suggest some very specific predictions. The first one is easy: there will never be any flying saucers on the White House lawn. If good luck in an immense universe has played the kind of critical role I've been suggesting, then advanced civilisations elsewhere are inevitable – but they will also be so far away that we will never be able to communicate with them or even observe influences they may have on their galactic neighbourhoods. We are certainly not close enough for little green men to drop in when they happen to be in the neighbourhood. However, the trouble with this prediction is that it's out of our hands. How long do we wait before realising that no visitors are coming?

What's really needed are predictions that we can actively go out and test for ourselves. A good place to start is by looking for life elsewhere in the solar system. My hope is that we will soon find microscopic life living beneath the surface of Mars and my expectation is that its biochemistry will show it to be similar to Earth life. This will generate some interesting discussions as we debate whether this is evidence that there is only one way to make life or evidence for cross-contamination between the worlds. I expect a consensus to eventually emerge that the similarities are too great to be explained by a separate origin and we will then know that 'Earth life' exists on two planets in our solar system. Even that is a very exciting prospect and one that I may be lucky enough to live to see. We'll then have to decide whether to make Mars off-limits so that its unique biosphere can develop without interference from mankind. I hope that we do.

The rest of the solar system is probably ours to exploit; I believe that the origin of life, like all the major steps leading to the emergence of intelligence, is a rare occurrence. Contamination of Europa or Enceladus by microbe-bearing meteorites from Earth is unlikely and an independent origin for life would therefore have to be taken very seriously if life is found in such places. An even more exciting prospect would be the discovery of life in the ethane lakes of Titan, because this would surely be so unlike Earth life that an independent origin would have to be the preferred explanation. However, if the thrust of this book is correct, even simple life is rare and my expectation is that the giant-planet moons of the solar system will turn out to be completely sterile. We can only tell by looking, though, and I really hope I'm wrong!

My final prediction is that the ongoing search for exoplanets will eventually show that the giant planets of our solar system are unusually far apart. If there really has been anthropic selection for an Earth with a stable axis, this implies that we live in a solar system whose orbits wobble unusually slowly, since the key to axial stability is avoidance of resonance, and this needs a gap between the relatively rapid axis-precession period of a planet and the slower periods at which orbits oscillate. This gap occurs only if major worlds are relatively small or relatively widely spaced so that orbits oscillate very gently. However, the gas giants in our solar system are not unusually light compared to known exoplanets, and so they must be more spread out than is usual. It will take many more decades of exoplanet searching to confirm this.

To be more specific, my expectation is that only about 1 per cent of planetary systems orbiting Sun-like stars will have a fundamental orbital oscillation period slower than the solar system's 50,000 years. This prediction can be tested using current technology and that brings me back to OGLE-2006-BLG-109L b and her sister world, the equally memorable OGLE-2006-BLG-109L c, mentioned in Chapter 3. OGLE stands for the optical gravitational lensing experiment, a Polish-run project to look for evidence of dark matter by monitoring the brightness of millions of stars in the galactic bulge (hence BLG in the planets' names) and elsewhere. OGLE has almost incidentally proved to be very successful at detecting planets from the months-long brightening that occurs when one star passes in front of another so that its gravity acts like a lens and focuses the light of the more distant star onto the Earth. Deviations from a simple pattern of steady brightening over several weeks followed by equally steady fading indicate the presence of planets around the nearer star. And, unlike the RV method I discussed in some detail earlier, this gravitational microlensing is sensitive to planets orbiting relatively far from their star. OGLE's 109th lensing event (hence 109L) of 2006 proved to be a particularly exciting example of planet detection by this technique. The results when analysed showed the presence of two worlds (hence 'b' and 'c'). Planet 109L b turned out to be a little smaller than Jupiter and orbiting its half-solar-mass star at about half the distance Jupiter does from our Sun. Its companion, 109L c, is slightly smaller than Saturn and has an orbit about half the diameter of Saturn's. So, OGLE-2006-BLG-109L looks like a solar system in miniature. This discovery of a mini solar system 5,000 light years from Earth demonstrates that we now have the technical ability to search for true analogues to the solar system, stars whose main companions are not hot Jupiters but gas giants in orbits comparable to the major planets of our own system.

OGLE-2006-BLG-109L is the most similar planetary system to ours yet found and, if solar-like systems are common, it should only be a matter of time before a truly comparable one is discovered. I think this search will fail though, because, as I said above, I believe that our planetary system is an unusually open one. In fact, no true solar system analogues have yet been found, although it will take many more decades before enough data has been collected to conclusively show that they are as rare as I'm suggesting.

There must be many other anthropic-based predictions concerning properties of extra-solar planetary systems that can be tested against data likely to become available as the 21st century progresses. If such predictions are successful, the increased credibility of anthropic ideas will profoundly change our understanding of the Earth and of our place in the Universe. My own motivation for beginning the research described in this book was that I felt that the anthropic principle, if true, should be one of the cornerstones of our understanding of geology. A good example of how anthropic reasoning can alter our view of important Earth processes concerns the issue of climate, which has been so central to my story. The observation that the Earth's temperature has been surprisingly consistent over billions of years seems to cry out for an explanation in terms of stabilising processes. However, if climate stability is the consequence of anthropic selection then there is no grand stabilising process at all and searches for it will, ultimately, fail. That's not to say that we shouldn't look for it, it's just that we shouldn't necessarily conclude that we're missing something if the search is unsuccessful.

Acceptance that the Earth is a very odd planet, and that this was necessary for the emergence of humans, also has a very obvious impact on the search for extraterrestrial intelligence. Quite bluntly, if there is significant anthropic selection for Earth properties, then we are probably effectively alone in the Universe. As I discussed earlier, the nearest extraterrestrial advanced civilisation could easily lie beyond the edge of the visible Universe and so be uncontactable. This is quite a disappointing conclusion for many. Indeed, one prominent, well-informed critic of anthropic ideas has admitted that his views may be coloured by having grown up watching the original *Star Trek* series. Maybe my own views have been coloured by slightly more recent films. I've thought for a long time that *Alien* was far more plausible than Mr Spock, so it's quite possible that my subconscious doesn't want aliens to exist.

There may be other psychological factors influencing people's rejection of anthropic ideas. In particular, the concept of a balance of nature is deeply ingrained in western, and other, cultures. The idea that this balance may, at least in part, be an illusion is quite hard for many people to swallow. One particularly dangerous example of unsubstantiated belief in inherent balance is the rejection, by some, of the idea that humans are beginning to adversely affect the Earth's climate. Opponents of the global warming hypothesis contend that some, as yet undiscovered, process will automatically compensate for mankind's interference in the atmospheric composition. This seems to be based entirely on a naive belief that the Earth's climate system is naturally stable, even though the geological evidence clearly shows that our climate has changed abruptly many times in the past; the last time just 11,000 years ago. Note that, if I'm right and the absence of really serious climate instability over the last half-billion years is just a fluke, we need to be very, very careful when we meddle with the atmosphere. We may just find out the hard way that planets with nasty climates are quite easy to produce.

Another psychological factor that, perhaps, makes anthropic selection hard to accept is that we are so used to our wonderful planet that we rarely stop to notice what an amazingly beautiful, unique and, dare I say it, miraculous place it is. If we didn't take the Earth quite so much for granted it would become obvious, I think, that it is an extremely odd place. Many people may find the view of the Universe set out in this book rather bleak, but for me, the anthropic principle has merely served to sharpen my appreciation of our stunning home-world and how lucky I am even to exist.

So, I certainly believe that the possibility that the Earth is special should be taken very seriously by everyone and for all sorts of reasons, but in conclusion, I'd like to finish with the most important justification of all for considering this idea. It's probably true.

Epilogue: Siblings

Half a billion years from now, the Earth has become a very harsh and unforgiving place. Carbon dioxide has disappeared from the atmosphere, photosynthesis has become impossible and plants are extinct. Oxygen, too, is now absent from the air. Average temperatures are approaching 60°C and noon-day equatorial temperatures now exceed boiling point. In the high Arctic a new mountain chain has grown that rivals the Himalayas of our own time, and towards the tops of its north-facing slopes, drizzle reluctantly condenses out of the humid atmosphere to collect in muddy, tepid pools that are the Earth's only remaining standing water. A few particularly tough *Loricifera* cling to life here by grazing on bacterial mats and absorbing hydrogen sulphide in place of oxygen. Their time too is drawing to a close as temperatures soar to levels unknown for 5 billion years. Liquid water has become a scarce commodity that will vanish completely in a few more million years, taking the last of Earth's living things with it.

For almost 5 billion years the Earth maintained a fairly stable temperature because, as the Sun gradually warmed through time, our atmosphere slowly lost its greenhouse gases. However, even a zero level of greenhouse warming is now not low enough to keep our planet cool under the gaze of an ever-warming Sun. To make matters worse, water vapour and methane are now building up rapidly in the atmosphere and a runaway greenhouse is beginning. The long history of good weather on our world is about to come to an end. Life of any kind will soon become impossible and the unique 5 billion-year story of life on Earth will reach its sad conclusion.

Our complex biosphere has outlived that of our sister planet, Nemesis, by less than a billion years. Life on Earth will not even last long enough to suffer the same fate as our sibling. All our water will evaporate into space a billion years before the Moon moves far enough away for our axis to become unstable. The goddess of undeserved good fortune exacts her price from all inhabited worlds but she deploys many different agents of destruction.

I will not finish on such a negative note. Earth and countless other inhabited worlds scattered thinly throughout an unimaginably immense multiverse have given rise to the fragile wonder of life. On Earth we have laughed, loved and wondered at the beauty of the world and the Universe around us. We are part of an extraordinary miracle and I, for one, feel very lucky.

Further Reading

I believe that many people, perhaps most, are interested in the big questions this book has touched on, questions such as 'Why are we here?' and 'Are we alone?' I personally also find these questions interesting because they are covered by areas of science that are filled by controversy and passionate argument; and I do enjoy a good argument. This is a rapidly moving field which, as I hope this book has shown, is currently experiencing a dramatic influx of hard data concerning both the very ancient history of life on Earth and the existence of strange worlds beyond our cosy solar system. At present, there is no scientific consensus about what this data is telling us and, inevitably, I've put my own cast on the subject. But I passionately believe that, whatever the topic, alternative views must always be listened to, and so I'd like to spend the last pages of *Lucky Planet* suggesting where you might go to explore a wider range of opinions.

Your first port of call should be James Kasting's excellent *How to Find a Habitable Planet* (Princeton University Press, 2010), which is, to a considerable extent, a riposte to the equally engaging *Rare Earth* (Copernicus Books, 2000) by Peter Ward and Donald Brownlee. *Rare Earth*, like my own book, supports the view that the Earth may be special but it takes a much broader look well beyond the issue of climate stability that I have concentrated on. *How to Find a Habitable Planet* takes very much the opposite view and presents many cogent arguments suggesting that life may, after all, be common in the Universe. I suggest you read both books and make your own mind up.

That covers what could be described as the main controversy I've looked at in this book, but in fact almost every chapter of *Lucky Planet* discusses issues that cause heated debate among the experts studying them. A good example of this concerns the history of Earth's climate itself. Did those snowball Earth episodes really happen or has there been massive over-interpretation of very limited data? What about the cooling trend over billions of years: Is it real or an artefact of poor data? To read more about the Earth's climate history you could not do better than read *The Goldilocks Planet* (Jan Zalasiewicz and Mark William, Oxford University Press, 2012). I can also thoroughly recommend *Frozen Earth: The Once and Future Story of Ice Ages* by Doug MacDougal (University of California Press, 1994). When it comes to Gaia and climate, you must read *On Gaia* (Toby Tyrrell, Princeton University Press, 2013), which is an in-depth critique of the Gaia hypothesis. For a more positive assessment take a look at *Revolutions that Made the Earth* by Tim Lenton and Andrew Watson (Oxford University Press, 2011), whose main subject is an unmatched exploration of the most important events in the evolution of life on Earth.

On the cosmological side of my story there are many excellent texts available, since this is a very well covered area in popular science books. My personal favourite is Leonard Susskind's *The Cosmic Landscape* (Hachette Book Group, 2006) but I'd also suggest Steven Manly's *Visions of the Multiverse* (The Career Press, 2011) and *The Goldilocks Enigma* by Paul Davies (Penguin Books, 2006), which are, perhaps, a little more accessible for those new to the subject. If, like me, you find that a little mathematics enlightens rather than obscures, then have a go with Barbara Ryden's *Introduction to Cosmology* (Addison Wesley, 2003), which, as a non-expert in this area, I found enormously helpful.

Finally, for the technical references I have used in my work, you will need to go to my published scientific papers on the subject. Links to these can be found at www.davidwaltham.com, where you should select the *Lucky Planet* menu.

Acknowledgements

I could not have written a book covering such a diverse range of topics without a great deal of help from many fellow scientists. The following experts read passages, entire chapters or the whole book for me and offered friendly advice and mildly-worded corrections: Margaret Collinson, Lewis Dartnell, Roberto Donovaro, Richard Ghail, Jonti Horner, Jim Kasting, Andrew Liddle, Euan Nisbet, Hiranya Peiris, Toby Tyrrell and Kathy Whaler. In addition, my colleague Peter Burgess read each chapter as I completed it, suggested many areas for improvement and then read the chapters again! This would have been a very different, and far less accurate, book without his input and without the help of all these colleagues. Any remaining factual errors are, of course, my own responsibility. I should also emphasise that the help I received does not in any way imply that these colleagues endorse what I have said in these pages. In fact, several of them strongly disagree with my views, but that just emphasises how wonderfully generous they have been with their time and advice. It's a pleasure to acknowledge, as well, the helpful encouragement and comments from my 'intelligent lay-people', my wife Joanne Waltham and my uncle Roger Waltham.

I must add how grateful I am for the unwavering support, encouragement and practical advice from my agent, Sally Holloway, over the several years that it took for this project to come to fruition. Finally, the staff at the publishing companies (Icon Books and Basic Books) were absolutely brilliant – thank you for the diplomatically expressed guidance on which parts needed a bit more work.

Index